Workshop on Sources of Polarized Leptons and High Brightness Electron Beams 2010

AA002814

(PESP 2010)

Journal of Physics: Conference Series Volume 298

Bonn, Germany
21 – 24 September 2010

Editors:

Wolfgang Hillert
Kurt Aulenbacher

ISBN: 978-1-61782-927-7
ISSN: 1742-6588

Printed from e-media with permission by:

Curran Associates, Inc.
57 Morehouse Lane
Red Hook, NY 12571

Some format issues inherent in the e-media version may also appear in this print version.

Printed by Curran Associates, Inc. (2011)

Institute of Physics
Dirac House, Temple Back
Bristol BS1 6BE UK

Phone: 44 1 17 929 7481
Fax: 44 1 17 920 0979

techtracking@iop.org

Additional copies of this publication are available from:

Curran Associates, Inc.
57 Morehouse Lane
Red Hook, NY 12571 USA
Phone: 845-758-0400
Fax: 845-758-2634
Email: curran@proceedings.com
Web: www.proceedings.com

TABLE OF CONTENTS

INTRODUCTION

POLARIZED GUNS

HIGH VOLTAGE DC GUNS

RF GUNS

PHOTOCATHODES

NEW APPROACHES

POLARIZED POSITRONS

POLARIMETRY

PREFACE

This volume contains the proceedings of the International Workshop on Sources of Polarized Leptons and High Brightness Electron Beams (PESP2010). The workshop was hosted by the University of Bonn from 21–24 September 2010. PESP2010 has been conducted as satellite workshop preceding the International SPIN Physics Symposium (SPIN2010). In view of the expanding and challenging field of the production of high brightness lepton beams, the selection of workshop topics was extended to cover the demands of modern accelerator developments and applications. As novel light sources such as FELs and ERLs place new requirements on the properties of unpolarized beams in terms of brightness and intensity, the generation of DC and pulsed beams was included as a major topic in the scientific program. In keeping with tradition, polarized beams formed another key topic. The discussion was however extended to the generation of polarized positron beams, as e.g. required for future linear colliders. Besides the presentation of recent developments in these fields and status reports from laboratories worldwide, additional emphasis was put on the applications of polarized electron beams.

PESP2010 was attended by 61 registered participants from all over the world. The scientific program comprised 30 talks in plenary sessions and 7 posters. An overview of the scientific results presented at PESP2010 was given in a summary talk at the SPIN2010 symposium. In total 24 contributions to the proceedings were submitted and reviewed by members of the scientific advisory committee before publication in this volume.

Many people contributed to the success of the workshop. Among them are members of the Local Organizing Committee as well as the scientific and technical staff of the Physics Institute and the in-house accelerator facility ELSA, who patiently and efficiently took care of the numerous organizational aspects related to the event. The environment they created was highly conducive to a productive working atmosphere and fruitful discussions. The dedication of all who contributed is gratefully acknowledged. We would like to extend our warmest thanks to all of them.

Bonn, May 10, 2011

Wolfgang Hillert and Kurt Aulenbacher
Editors

Journal of Physics: Conference Series **298** (2011) 012001 doi:10.1088/1742-6596/298/1/012001

Impact of polarized e^- and e^+ beams at a future Linear Collider and a Z-factory

Part II – Physics beyond the Standard Model

Gudrid Moortgat-Pick

University of Hamburg, Luruper Chaussee 149, 22761 Hamburg and DESY, Notkestrasse 85, 22607 Hamburg

E-mail: gudrid.moortgat-pick@desy.de

Abstract. Polarization of both beams at a future Linear Collider would be ideal for facing both expected and unforeseen challenges in searches for new physics: fixing the chirality of the couplings and enabling the higher precision for the polarization measurement itself as well as for polarization-dependent observables, it provides a powerful tool for studying new physics at the future Linear Collider, such as discovering new particles, analyzing signals model-independently and resolving precisely the underlying model. Techniques and engineering designs for a polarized–positron source are well advanced. Potential constraints concerning luminosity, commissioning and operating issues appear to be under control. This article mainly treats with the impact of polarized beams on physics beyond the Standard Model, whereas the fundamentals in polarization as well as the gain in electroweak precision physics are summarized in the corresponding part I [1].

1. Introduction

The first exploration of the TeV energy scale is expected to be made with proton–proton collisions at the Large Hadron Collider (LHC). Its discoveries would be complemented by results from the electron–positron International Linear Collider (ILC), that is currently already in a mature technical design phase and by results from a multi-TeV option of a linear collider (LC), CLIC, awaiting its prototype feasibility experiments in the next years. It is expected that the clean signatures and in particular the precise measurements made possible by a high luminosity LC at a known and tunable beam energy will bring revolutionary new insights into our understanding of the fundamental interactions of nature and the structure of matter, space and time. In the hunt for physics beyond the Standard Model (SM), only small signs may be visible, and the ILC provides already optimal conditions for searching for the unexpected.

The physics return from the investment in the linear collider would be maximized by providing polarized electron and positron beams. It is recognized that beam polarization can play an important role in the ILC programme, and polarization of the electron beam is already foreseen for the baseline design [2]. A high degree of at least 80% polarization is already envisaged, and new results indicate that even 90% may be achievable. A polarized electron beam would already provide a valuable tool for scrutinizing the SM and diagnosing new physics.

The possibility of polarizing the positron beam is currently discussed as one upgrade option for the ILC. In the report [3] it is shown that the full potential of the ILC will be realized only with a polarized positron beam together with a polarized electron beam. In addition to detailed studies of the SM and properties of new particles, as well as new kinds of interactions, the polarization of both beams would also enable indirect searches with high sensitivity for new physics in a largely model-independent approach. In the following a short summary about the impact of beam polarization in new physics searches is given, whereas in part I, [1], the focus is on the impact of beam polarization for electroweak high precision studies. This article closes with a short overview of technical issues and their current status.

2. The physics case for polarized positrons

2.1. Direct searches for new physics with longitudinally-polarized beams

The dominant processes in e^+e^- experiments are annihilation (s-channel) and scattering (t-channel) processes. In t-channel processes the helicities of the electrons and positrons can be related directly to the chirality and properties of the (new) particles produced. In annihilation processes the helicities of the electron and positron are related by the spin of the particle(s) exchanged in the s-channel. Suitable combinations of the electron and positron beam polarizations may be used to enhance significantly signal rates and also to suppress efficiently unwanted background processes. These capabilities are particularly welcome in outlining searches for new physics, where in many cases only very small rates are predicted. An increased signal/background ratio combined with high luminosity provides a promising environment for discoveries even at the edge of the kinematical reach.

If both beams are polarized, in t-channel processes the helicities/chiralities of the electron and positron can be related directly to the properties of any produced (new) particles and its interactions. The ability to adjust independently the polarizations of both beams simultaneously provides unique possibilities for probing directly the properties of the produced particles. In particular, it becomes possible to access directly their quantum numbers and chiral couplings, with a minimal number of assumptions.

2.1.1. Examples in searches for Supersymmetry One of the best motivated extensions of the SM is Supersymmetry (SUSY). This theory predicts that all new SUSY particles carry the same quantum numbers as their SM partner particles, with the exception of the spin, which differs by half a unit. A prominent sector is represented by partners of the left– and right–chiral electrons/positrons, the scalar electrons/positrons $\tilde{e}^{\pm}_{L,R}$, which have to be associated to their SM-partners.

To probe their quantum numbers one has to separate experimentally the pairs $\tilde{e}^+_L\tilde{e}^-_R$ produced only by a t-channel process from the pair $\tilde{e}^+_R\tilde{e}^-_R$ produced by an s-channel process. In quite a number of scenarios even a highly polarized electron beam will not be sufficient to separate such pairs, because both pairs are produced with almost identical cross sections and have the same decays, see Fig. 1 (lower plot). With polarized positrons in addition to polarized electrons, the pairs have different cross sections and the $\tilde{e}^+_L$ and $\tilde{e}^-_R$ can be distinguished by charge conjugate separation, Fig. 1(upper plot). As seen from this example, polarized positrons may be essential when probing properties of new physics.

The polarization of both beams allows one to probe directly not only the chiral quantum numbers as shown in Fig. 1, but also the spins of particles produced in resonances.

A prominent example in an R–parity violating SUSY model is the production of a spin-0 particle, the scalar neutrino, with $\mu^+\mu^-$ in the final state. Since the sneutrino couples only to left-handed $e^{\pm}$, the peak is strongest for the LL polarization configuration, a signature that would point directly to the presence of a spin-0 resonance, Fig. 2. The SM background is strongly suppressed and one gets a $S/B \sim 11$ for $(P_{e^-}, P_{e^+}) = (-80\%, -60\%)$, whereas for $(P_{e^-}, P_{e^+}) = (-80\%, 0\%)$ the ratio is only $S/B \sim 4$. Conversely, in the case of a spin-1 resonance, e.g. the Z' particle in the SSM model, see Fig. 3, the corresponding resonance peak would be strongest for the LR configuration, with a similar polarization dependence as the SM background. This simple example shows how one can disentangle the form of interaction if both beams are polarized.

2.2. Indirect searches for new physics with longitudinally-polarized beams
Some new physics scales, such as those characterizing gravity in models with extra dimensions or the compositeness scale of quarks and leptons, could be too large to be directly accessible at the energies of present as well as future accelerators. Therefore it will also be important at the ILC to devise indirect search strategies for new physics, with high sensitivity and large model-independence. Indeed, the ILC has a large discovery potential for indirect searches beyond the kinematical limit. Effective interactions represent a general framework for the low-energy parametrization of the effects of non-standard dynamics due to exchanges between SM particles of very heavy states with masses beyond the available accelerator energy. This is the case of the four-fermion contact interactions (CI) inspired by compositeness but applicable much more generally, and of the mass scales characterizing models of gravity in large extra dimensions. Manifestations of such new interactions can be probed through deviations of cross sections from the SM predictions, and indirect bounds on the new energy scales and coupling constants can thereby be derived.

Longitudinal polarization of both beams is decisive for deriving model-independent bounds on the different possible couplings. With both beams polarized, the error in $\Delta P/P$ is reduced, the accuracy of the $A_{\rm LR}$ measurement is considerably enhanced, for instance, an polarization degree of $P_{e^+} = 60\%$ reduces the relative error $\Delta A_{\rm LR}/A_{\rm LR}$ by more than a factor 3. Furthermore more observables can be defined, which enables one to disentangle and constrain the different couplings in a model-independent approach.

2.2.1. Example from contact interactions For example, in Bhabha scattering the four–fermion contact interactions (CI) are parametrized by three couplings ($\epsilon_{\rm RR}$, $\epsilon_{\rm LR}$, $\epsilon_{\rm LL}$). The t–channel contributions depend only on $\epsilon_{\rm LR} = \epsilon_{\rm RL}$, whereas the s–channel contribution depends only on the pairs ($\epsilon_{\rm RR}$, $\epsilon_{\rm LR}$), ($\epsilon_{\rm LR}$, $\epsilon_{\rm LL}$). In order to derive model–independent bounds, it is necessary to have both beams polarized. Tight bounds up to 5×10^{-4} TeV^{-2} at 95% CL can be derived. It has been assumed that no deviations from the SM are measured within the experimental uncertainty in the observables, i. e. the combinations of polarized cross sections σ_{++}, σ_{+-} and σ_{-+}, see Fig. 4.

2.2.2. Example from Z' model Extra neutral gauge boson Z' can be probed by their virtual effects on cross sections and asymmetries. For energies below a Z' resonance, measurements of fermion–pair production are sensitive only to the ratio of Z' couplings and Z' mass. As an example, beam polarizations $(P_{e^-}, P_{e^+}) = (80\%, 60\%)$ would improve the measurement of the $b\bar{b}$ couplings of the Z' without knowledge of the Z' mass by about a factor 1.5, compared to $P_{e^-} = 80\%$ only, see Fig. 5. The crucial point is the fact that the systematic uncertainties can be significantly reduced when both beams are polarized.

2.3. Searches for new physics with transversely-polarized beams
With both beams polarized, another powerful tool would be available at the ILC, namely the use of transversely-polarized beams, which enhances the physics potential significantly in SM physics as well as in different new physics models. However, both beams have to be polarized, otherwise all effects at the leading order from transverse polarization vanish for $m_e \to 0$ (suppression by $m_e/\sqrt{s}$).

New CP–sensitive observables can be constructed in general, and azimuthal asymmetries can be exploited. These new observables are important e. g. in SUSY searches for the resolution

Journal of Physics: Conference Series **298** (2011) 012001 doi:10.1088/1742-6596/298/1/012001

of new CP–violating phenomena. They further enlarge the number of observables available to constrain the new physics parameters. Since e.g. in SUSY many of the 105 free parameters are possible CP-violation phases, such tools may become particularly important in the direct searches for SUSY.

2.3.1. Example from extra dimensions In that context, the use of transversely-polarized beams offers new observables to detect non-standard interactions which are not of the current-current type, such as those mediated by spin-2 gravitons or (pseudo)scalar exchanges even in indirect searches. Sensitivities to a high mass scale of e.g. the extra dimension model up to ≥ 3 TeV are achievable, enabling even a model distinction.

The success to identify new physics even in indirect searches via polarized e^- and e^+ beams would represent a step forward of utmost importance for our understanding of fundamental interactions.

One representative example is the distinction between extra dimensions in the models of Randall–Sundrum (RS) and Arkani–Hamed, Dimopoulos, Dvali (ADD). With transversely–polarized beams a new asymmetry in $\sin 2\phi$ can be constructed, which is sensitive to the cut–off independent imaginary parts of the amplitude originating from the exchange of the (almost) continuous spectrum of ADD gravitons. Below the graviton resonance poles no imaginary parts emerge in the RS model, if one neglects the (small) widths with respect to the masses. The new asymmetry therefore vanishes for both the SM and the RS scenario, so that a non–zero value unambiguously signals the ADD graviton exchange, see Fig. 6. Such a model distinction is achievable up to ≥ 3 TeV.

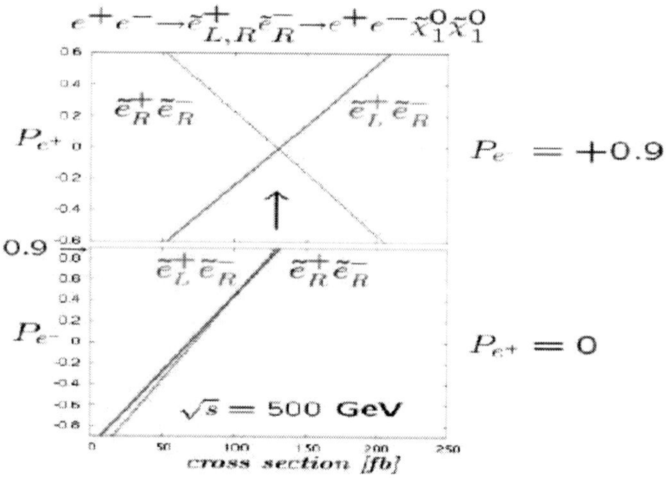

Figure 1. Separation of the selectron pair $\tilde{e}^+_L\tilde{e}^-_R$ in $e^+e^- \to \tilde{e}^+_{L,R}\tilde{e}^-_{L,R} \to e^+e^-2\tilde{\chi}^0_1$ is not possible with electron polarization only (lower panel). If, however, both beams are polarized, the cross sections (upper panel) differ and the RR configuration, e. g. $(P_{e^-}, P_{e^+}) = (+90\%, +60\%)$ [3] clearly separates the pair $\tilde{e}^+_L\tilde{e}^-_R$. The SUSY parameters are chosen as in scenario S1 in [3].

3. Technical issues

3.1. Status of the Linear Collider

Two accelerator technologies are discussed for a linear collider.

a) In the energy range of $\sqrt{s} = 0.5$–1 TeV the superconducting technology, as implemented in the International Linear Collider (ILC), is the mature concept [7] to provide the expected unique scientific opportunity and enter a new precision frontier. The Reference Design Report (RDR) has been finished in 2007 [8]. No technical obstacles are predicted for the ILC design, but careful studies of possible cost saving changes for the current design have

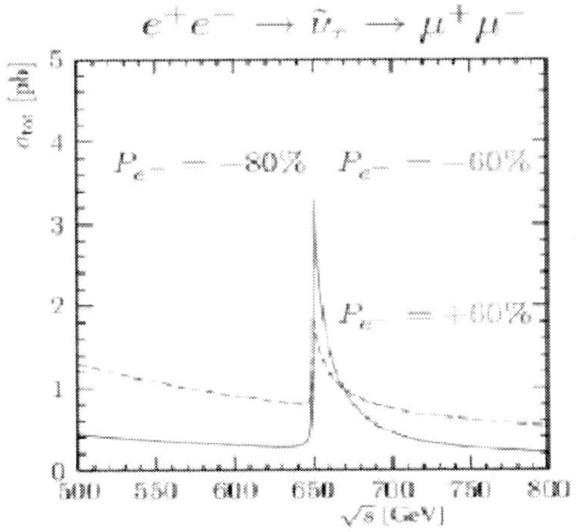

Figure 2. Sneutrino production in the R-parity-violating model: resonance production for $e^+e^- \to \tilde{\nu}_\tau \to \mu^+\mu^-$ (left panel) [3] for different configurations of beam polarization: $(P_{e^-}, P_{e^+}) = (-80\%, +60\%)$ (dashed), $(-80\%, -60\%)$ (solid). The striking effects of the spin-0 s-channel exchange can be enhanced using the LL configuration of beam polarization. The study was made at $\sqrt{s} =$ 650 GeV for $m_{\tilde{\nu}} = 650$ GeV, $\Gamma_{\tilde{\nu}} = 1$ GeV and the R-parity-violating couplings $\lambda_{131} = 0.05$ and $\lambda_{232} = 0.05$.

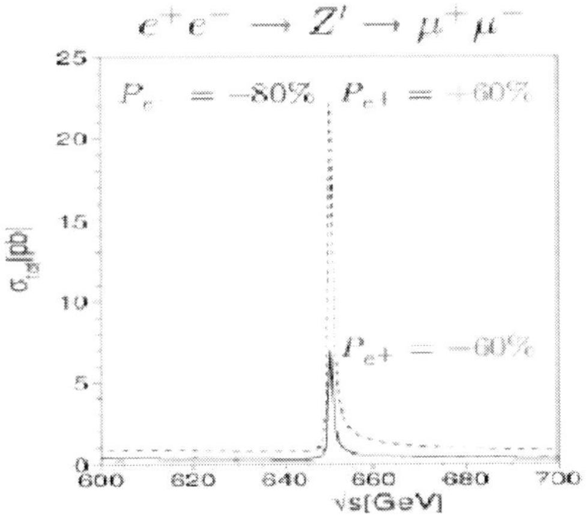

Figure 3. Z' production in the SSM model with a Z'-coupling of about 0.2, $m_{Z'} =$ 650 GeV and a very small width of about ≤ 1 GeV: resonance production for $e^+e^- \to Z' \to \mu^+\mu^-$ (right panel) [3]. The strong peak enhancement for the LR configuration with $(P_{e^-}, P_{e^+}) = (-80\%, +60\%)$ (dashed) is clearly dominant, whereas for the LL configuration $(P_{e^-}, P_{e^+}) = (-80\%, -60\%)$ (solid) —although not completely suppressed, due to only partial polarization— the signal is much smaller.

to be done with regard to their possible impact on the physics potential of the machine. The Technical Design Phase (TDP) of the ILC is under the responsibility of the 'Global Design Effort' (GDE).
Starting the industrial engineering phase, the optimization of, for instance, the cavities shapes are under study. Higher gradients up to 59 MV/m have already been achieved in single cells with so-called 're-entrant' cavities, developed by Cornell and KEK. Concerning the industrial cavity production, an average of 36 MV/m in nine-cell cavities has been achieved [9].

b) For achieving higher energies of $\sqrt{s}$ in the multi-TeV range a normal conducting two-beam acceleration concept is discussed, the Compact Linear Collider (CLIC). A conceptual design report is foreseen for 2010, where the key feasibility issues of the CLIC technology are foreseen to be demonstrated as well as the preliminary performance and a first cost estimation.

A fruitful ILC/CLIC collaboration has been started to address common R&D issues in the civil engineering & conventional facilities, beam delivery system, beam dynamics and detectors, more details see [9]. Comparing the potential of the two linear collider technologies several technical issues have to be taken into account that may have impact on the physics potential. Many precision measurements at a linear collider depend crucially on machine parameter more than on the achievable detector precision. For instance, the average energy loss i.e. beamstrahlung, has impact on the precision achievable via threshold scans and (polarized) cross sections. Beamstrahlung is predicted to be 2.4% at ILC with $\sqrt{s} = 500$ GeV (ILC500),

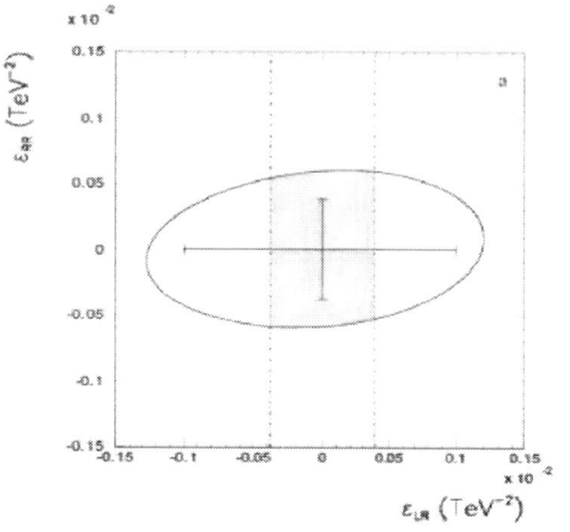

Figure 4. Contact-interaction in Bhabha scattering process $e^+e^- \to e^+e^-$: Allowed areas at 95% C.L. in the plane $(\epsilon_{\rm LR}, \epsilon_{\rm RR})$. The study was done at $\sqrt{s} = 0.5$ TeV, $\mathcal{L}_{\rm int}(e^+e^-) = 50$ fb^{-1}, $(|P_{e^-}|, |P_{e^+}|) = (80\%, 60\%)$. Vertical dashed lines indicate the range allowed for $\epsilon_{\rm LR}$ by $\hat{\sigma}_{{\rm LR},t}$. The cross indicates the constraints obtained by taking only one non-zero parameter at a time instead of two simultaneously non-zero and independent parameters [4].

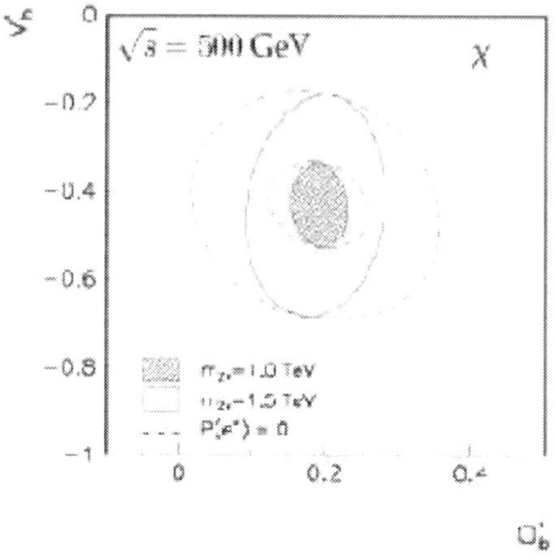

Figure 5. Determination of Z' couplings and masses with polarized beams: 95% C.L. contours for the axial (a_b') and vector (v_b') couplings of the Z' for $M_{Z'} = 1.0$, 1.5 TeV in the χ realization of an E_6-model [5] with $\sqrt{s} = 500$ GeV and $\mathcal{L}_{\rm int} = 500$ fb^{-1}. The solid line corresponds to $P_{e^+} = 60\%$, the dashed lines to $P_{e^+} = 0$ [5]; a_b', v_b' denote the couplings of Z' to the b-quarks.

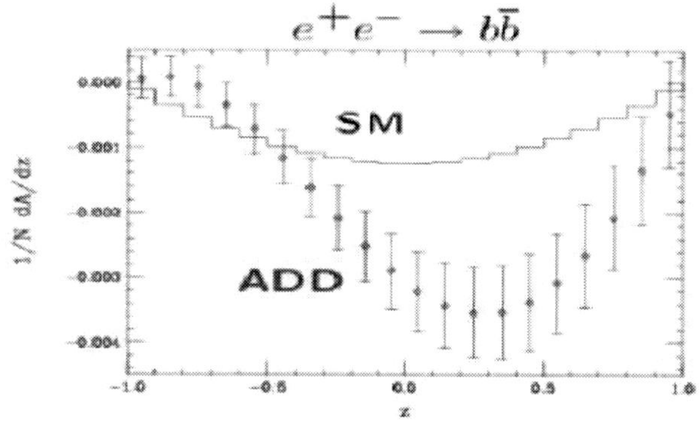

Figure 6. Differential azimuthal asymmetry distribution for $e^+e^- \to f\bar{f}$, e.g. $b\bar{b}$, at a 500 GeV LC assuming a luminosity of 500 fb^{-1}, $z = \cos\theta$. The histograms are the SM predictions while the data points assume the ADD model with $M_H = 1.5$ TeV; $(P_{e^-}^{\rm T}, P_{e^+}^{\rm T}) = (80\%, 60\%)$ [6].

7% at CLIC with $\sqrt{s} = 500$ GeV (CLIC500) and 29% at CLIC technology with $\sqrt{s} = 3000$ (CLIC3000). A formidable experimental challenge arises from the short (0.5 ns) bunch spacing at CLIC. Severe impact on the achievable precision due to pile-up of soft hadronic interactions can arise unless unprecedented time-stamping capability both for charged and neutral particles can be implemented into the CLIC detectors. Detailed simulations will be needed for achieving conclusive results concerning the physics potential of the different designs. Therefore a staged approach between the different designs may be beneficial [9].

3.2. Polarized positron sources

The two main methods for positron production under consideration for the ILC are a photon-based source and a 'conventional' source. The photon-based source uses multi-MeV photons and relatively thin targets (less than a radiation length thick) to produce positrons. If the photons are circularly polarized, the positrons (and electrons) are spin polarized. This positron polarization can be preserved in the subsequent capture, acceleration, damping, and transport to the collision point(s). The conventional source uses a multi-GeV electron drive beam in conjunction with thick, high-Z targets to produce positrons from the resultant electromagnetic cascade in the target. The positrons produced by this method cannot be polarized. These two schemes ultimately present very similar engineering challenges while at the same time having distinct attributes and drawbacks. It is emphasized, however, that only the photon-based schemes offer the promise of positron polarization.

Circularly-polarized photons are required for the generation of longitudinally-polarized positrons via $e^{\pm}$ pair production in a thin target. The photons are in the energy range of a few MeV up to about 100 MeV. Because the target is typically a fraction of a radiation length thick, high strength materials, such as titanium alloys, can be considered as opposed to conventional targets, where high-Z, high-density materials are required to minimize the emittance of the produced positrons. The two methods for generating the polarized photons under consideration are:

- a high-energy electron beam ($\geq$150 GeV) passing through a short period, helical undulator. The E-166 experiment at SLAC, was a demonstration of this undulator-based polarized positron production scheme with great success [10].
- Compton backscattering of laser light off a GeV energy-range electron beam. The concept is being tested in an experiment which is currently running at KEK [11].

In both schemes a positron polarization of about $|P_{e^+}| \geq 60\%$ is expected at the ILC. It is expected to reach a polarization degree of $P_{e^+} \sim 60\%$ without any loss of luminosity, higher polarization seems to be achievable at cost of luminosity. But already within the baseline confguration a polarization of the positron beam of about 30% may be achievable [12].

Both schemes would be applicable for the ILC design [2] and also adaptable for a possible future multi-TeV LC design, CLIC.

3.3. Polarimetry requirements

In order to fully exploit the polarization of the beams, one also has to measure precisely the actual degree of polarization. Therefore high precision polarimetry is mandatory. At the SLC one achieved already a precision of $\Delta P(e^-)/P(e^-) \sim 0.5\%$ with Compton polarimetry measured via a magnetic spectrometer. The goals at the ILC are even more challenging and one aims for $\Delta P(e^{\pm})/P(e^{\pm}) \leq 0.25\%$. In order to achieve such a precision, Compton polarimeters in combination with a dedicated chicane system and Cerenkov detectors are implemented as upstream polarimeter. A downstream polarimeter is further required and is applicable due to the crossing angle. Further details, see [13]. Such a dual measurement enables machine feedback and provides access to a precise determination of the luminosity-weighted polarization at the interaction point if precise spin tracking is provided, see [14].

The main depolarization effects at a LC are the following two effects: the classical spin precession that is described via the Thomas–Bargmann-Michel-Telegdi equation (T-BMT) and the quantum-mechanical spin-flip process (Sokolov-Ternov effect). The largest effects are predicted for the beam-beam interaction region due to the strong field of the oncoming beam. Due such a strong field environment, higher-order quantum effects have to be calculated and taken into account as well, further details see [15]. The resulting depolarization effects have been evaluated and compared for the ILC RDR and the current CLIC design, see Table 1. Smaller depolarization effects are expected to occur in the damping rings, the spin rotators and the beam delivery system, but they have to be included. This work is still ongoing.

4. Conclusions

To face the (expected and unforeseen) challenges of possible new physics followed by a precise revealing of the underlying model requires the polarization of both beams as a superior experimental tool. It enables to probe directly new physics properties, to detect maybe even tiny traces of new CP-violating sources, to handle possible background processes from e.g. new physics itself. Furthermore it leads to higher statistics, better control of systematics and higher precision in polarization dependent observables as left-right asymmetries. Polarization of both beams at the ILC are therefore an ideal preparation for 'the unexpected' and it might be advisable to use this powerful and variable option already from the beginning. An upgrade of the machine and the complete implementation of another polarized-positron source at a later stage would require again a careful commissioning of a new source which causes in any case losses in the total time of running. Therefore in case that a polarized source is already feasible at the start of the ILC design and no major differences concerning the reliability and commissioning of the polarized-positron source compared to the conventional positron source is expected, the implementation of the polarized-positron source should at least be completely designed and is foreseen for the baseline set-up.

Table 1. Depolarization effects in the beam-beam interaction region for ILC with fully polarized beams (100/100), with partially polarized beams (80/30) and for the CLIC-G design [16].

Effect	Depolarization ΔP_{lw}		
Design	ILC 100/100	ILC 80/30	CLIC-G
T-BMT (spin precession)	0.17%	0.14%	0.10%
ST (spin flip process)	0.05%	0.03%	3.4%
incoherent pairs	0.0%	0.0%	0.06%
coherent pairs	0.0%	0.0%	1.3%
total	0.22%	0.17%	4.8%

Acknowledgments

The author would like thank the organzisers of the PESP2010 conference for the welcoming and inspiring atmosphere and in particular Dr. W. Hillert for his enthusiastic engagement and many interesting discussions.

References

[1] G. Moortgat-Pick, *Proc. of 'SPIN2010 – 19th International Spin Physics Symposium', October, 2010, Jülich, Germany.*

[2] P. Grannis, R. Heuer, S. Komamiya, M. Oreglia, F. Richard, D. Son, ILCSC Parameter Subcommittee, Parameters for the Linear Collider, Nov. 2006 and Sep. 2003.

[3] G. Moortgat-Pick et al., *Phys. Rep.***460** *(2008), 131 (Preprint hep-ph/0507011).*

[4] A. A. Pankov and N. Paver, *Eur. Phys. J. C* **29** *(2003) 313 (Preprint hep-ph/0209058).*

[5] R. Casalbuoni, S. De Curtis, D. Dominici, R. Gatto and S. Riemann, *Preprint LC-TH-2000-006, hep-ph/0001215*; A. Leike and S. Riemann, *Z. Phys. C* **75** *(1997) 341 (Preprint hep-ph/9607306).*

[6] T. G. Rizzo, *JHEP* **0302**, *008 (2003) (Preprint hep-ph/0211374)*; *JHEP* **0308**, *051 (2003) (Preprint hep-ph/0306283)*.
[7] J. E. Augustin et al., *Linear Collider, Final International Technology Recommendation Panel report*, 2004.
[8] J. Brau et al. [ILC Collaboration], *ILC Reference Design Report Volume 1 - Executive Summary, (Preprint physics.acc-ph/0712.1950)*.
[9] A. De Roeck et al., *Eur. Phys. J. C* **66**, *525 (2010) (Preprint hep-ph/0909.3240)*.
[10] G. Alexander et al., *Nucl. Instrum. Meth. A* **610**, *451 (2009) (Preprint physics.ins-det/0905.3066)*.
[11] T. Omori, M. Fukuda, T. Hirose, Y. Kurihara, R. Kuroda, M. Nomura, A. Ohashi, T. Okugi, K. Sakaue, T. Saito, J. Urakawa, M. Washio, and I. Yamazaki, *Phys. Rev. Lett.* **96** *(2006) 114801 (Preprint hep-ex/0508026)*.
[12] B. C. Barish, *ICFA Beam Dyn. Newslett.* **40**, *8 (2006)*.
[13] D. Käfer, *Proc. of 'SPIN2010 – 19th International Spin Physics Symposium', October, 2010, Jülich, Germany*.
[14] M. Beckmann, *Proc. of 'SPIN2010 – 19th International Spin Physics Symposium', October, 2010, Jülich, Germany*.
[15] A. Hartin, *Proc. of 'SPIN2010 – 19th International Spin Physics Symposium', October, 2010, Jülich, Germany*.
[16] I.R. Bailey, A.F. Hartin, G. A. Moortgat-Pick et al., *Proc. of 11th European Particle Accelerator Conference (EPAC 08), June 2008, Magazzini del Cotone, Genoa, Italy*.

Status and recent developments at the polarized-electron injector of the superconducting Darmstadt electron linear accelerator S-DALINAC

Y Poltoratska[1], C Eckardt[1], W Ackermann[2], K Aulenbacher[3], T Bahlo[1], R Barday[1], M Brunken[1], C Burandt[1], R Eichhorn[1], J Enders[1], M Espig[1], S Franke[2], C Ingenhaag[1], J Lindemann[1], W F O Müller[2], M Platz[1], M Roth[1], F Schneider[1], M Wagner[1], A Weber[1], T Weiland[2] and B Zwicker[1]

[1] Institut für Kernphysik, Technische Universität Darmstadt, D-64289 Darmstadt, Germany
[2] Institut für Theorie Elektromagnetischer Felder, Technische Universität Darmstadt, D-64289 Darmstadt, Germany
[3] Institut für Kernphysik, Johannes-Gutenberg-Universität Mainz, D-55218 Mainz, Germany

E-mail: `enders@ikp.tu-darmstadt.de`

Abstract. At the superconducting Darmstadt electron linac a 100 keV source of polarized electrons has been installed. Major components had been tested prior to installation at an offline teststand. Commissioning of the new source at the S-DALINAC will take place early in 2011. We report on the performance of the teststand, simulations, developments on the laser systems, new radio-frequency components for the S-DALINAC injector, and the status of the implementation of the source.

1. Introduction

The superconducting Darmstadt electron linear accelerator S-DALINAC [1] provides electron beams mainly for nuclear structure experiments with energies between 2 and 130 MeV since the early 1990s. Figure 1 shows an overview of the facility with the accelerator and the experimental installations. Behind the superconducting injector linac, experiments with bremsstrahlung are carried out for endpoint energies up to 10 MeV [2]. The main accelerator section is designed for an energy gain of up to 40 MeV, and two recirculating beam lines allow the main accelerator to be passed up to three times.

In the experimental hall a new photon tagger for nuclear structure experiments with photon energies of 5 – 20 MeV (electron energies of 20 – 30 MeV) has been installed [3] as well as a bremsstrahlung setup investigating the polarizability of the nucleon [4]. The workhorses for electron scattering experiments are the two magnetic spectrometers: At a large-solid-angle QClam-type spectrometer coincidence experiments are carried out (see, *e.g.*, [5]) as well as single-arm experiments at 180° (*e.g.*, [6]). A high-resolution magnetic spectrometer is used for inclusive electron scattering experiments (see, *e.g.*, [7]).

So far, the experiments utilize an unpolarized beam produced by a thermionic electron source. However, polarization degrees of freedom can provide access to further observables. Proposed

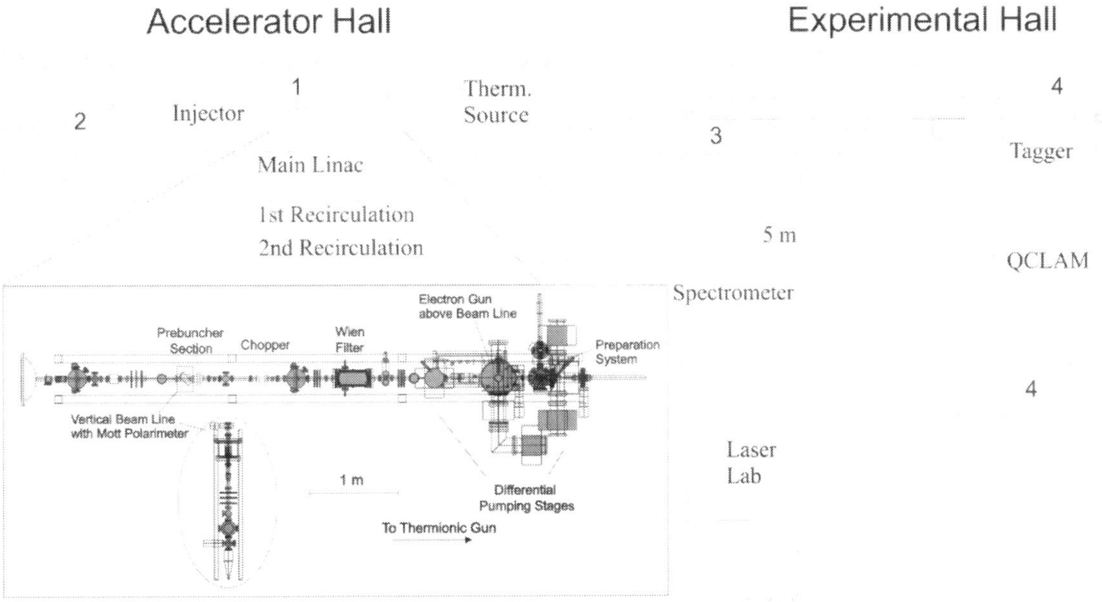

Figure 1. Overview over the superconducting Darmstadt electron linear accelerator S-DALINAC. The polarized source is shown schematically in the inset on the lower left. The polarized source is installed between the thermionic source and the superconducting injector linac and is driven by a laser beam transferred to the source from the laser laboratory. Polarimeters are being constructed for various experimental sites: (1) 100 keV Mott polarimeter; (2) Mott polarimeter and Compton-transmission polarimeter for 5-10 MeV experiments with bremsstrahlung; (3) location of a future 50-130 MeV Møller polarimeter; (4) Compton transmission polarimeters at the tagger system and the QClam magnetic spectrometer.

experiments comprise polarization correlation experiments in bremsstrahlung production (see references [8, 9] for examples), parity non-conservation experiments with regard to photo-induced fission, and, *e.g.*, the measurement of the fifth structure function in electron scattering in light nuclei. A brief account of the latter two experiments' status is summarized in a conference report [10].

As spatial dimensions are limited at the S-DALINAC, a compact source of polarized electrons needed to be constructed. The only suitable location was found to be the normal conducting injection section between the existing thermionic gun – which will remain in use – and the injector linac cryostat. This is indicated in figure 1. The basic layout started from an adaptation of the polarized electron source at the Mainz Microtrons MAMI [11]. The S-DALINAC polarized injector also uses photo-emission of electrons from irradiating a strained-superlattice GaAs cathode. The operation voltage amounts to -100 kV. Prior to installation of the source at the S-DALINAC, the main components have been set up at a separate teststand [12], and the performance of elements of the setup were analyzed in numerical simulations [13].

2. Teststand operation and performance

A schematic overview of the teststand is seen in figure 2. The electron beam is produced from the strained-superlattice cathode held at the tip of an electrode at the high-voltage potential of -100 kV. Separate preparation and load-lock systems provide for the (re-)activation and the fast exchange of cathodes, respectively. At the teststand, a diode laser system was used to irradiate the cathode. Preparation chamber, high-voltage chamber, and the adjacent vertical beam line including a 270° alpha magnet deflecting the beam into the horizontal direction form an ultra-high vacuum (UHV) system with typical pressure below $1.6 \cdot 10^{-11}$ mbar, achieved by ion-getter pumps and non-evaporable getter material. Typical cathode lifetimes, (*i.e.*, the time

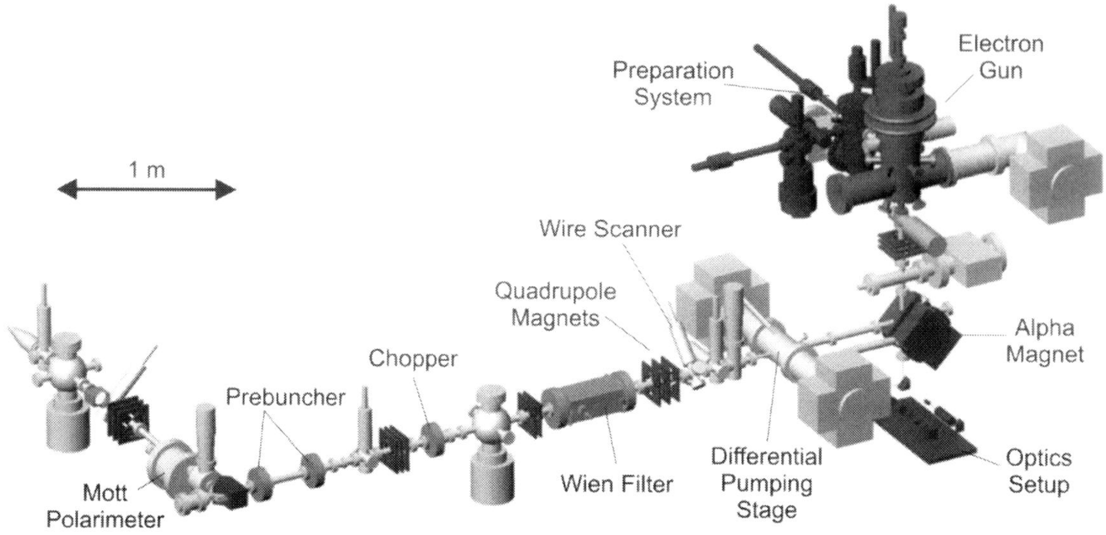

Figure 2. Technical drawing of the teststand of the source of polarized electrons. Electrons are released from a strained-superlattice cathode held at −100 kV in the cathode chamber. The cathode is prepared in an adjacent preparation system. After irradiation of the cathode with a laser beam from below, the electrons are released and guided downward. An alpha magnet deflects the beam into the horizontal direction. Beam transport is realized using short quadrupole triplets. Spin rotation is achieved by a Wien filter. A Mott polarimeter determines the degree of transverse polarization, and wire scanners and fluorescent screens allow for beam profile measurement.

in which the quantum efficiency of the cathode decreases to $1/e$ of its initial value, amounted to about 100 hours, and a maximum current of above 50 μA has been extracted so far.

In the horizontal test beam line, the polarized electrons leave the UHV system through a differential pumping stage. The electron spin direction can be rotated in the horizontal plane through a Wien filter, and very short quadrupole magnets (10 mm long, 52 mm aperture) are used for focusing of the beam. We note that a challenge in the numerical modeling of the quadrupoles is the handling of overlapping fields that occur since the quadrupoles are grouped in triplets [14].

Beam profiles are measured using fluorescent screens and wire scanners. A double solenoid together with a wire scanner serves as a setup for emittance measurement. Measuring the beam profile as a function of the solenoid's focusing strength allowed us to determine a normalized transverse emittance of the beam. Typical values of $\varepsilon_{n,x} = 0.15(3)$ mm mrad and $\varepsilon_{n,y} = 0.10(2)$ mm mrad were obtained for beam currents between a few nA and about 10 μA (DC operation). The results were – within the experimental uncertainties – very similar for bulk GaAs cathodes and strained-superlattice (high polarization) cathodes.

A Mott polarimeter employing several thin gold foils of varying thickness and housing four silicon surface-barrier detectors at 120° with respect to the incident beam is used for the determination of the degree of polarization. Polarization has been determined for different cathode types and laser wavelengths obtained from different diode laser systems. The maximum degree of polarization measured so far amounted to 86(3)% from a strained-superlattice cathode irradiated by a laser with 830 nm wavelength. Figure 3 depicts measured spectra of two different detectors in the Mott polarimeter that are located opposite of each other. The solid and dashed curves show the measured elastic electron scattering spectra depending on the helicity of the

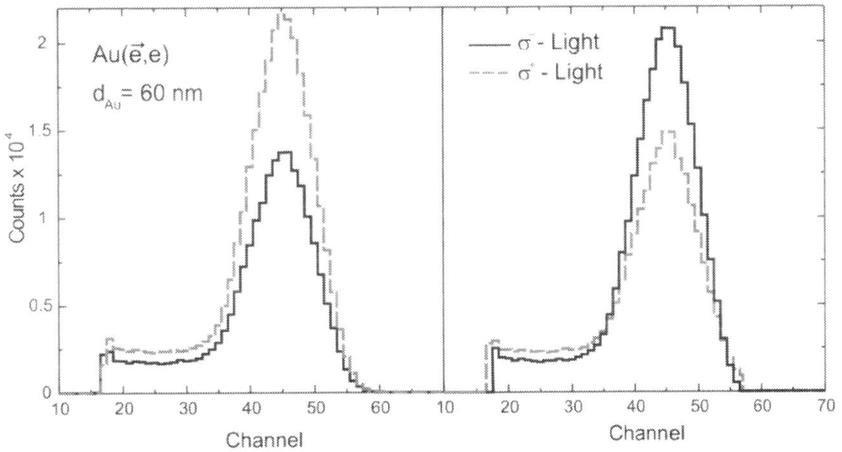

Figure 3. Measured elastic electron scattering spectra of transversely polarized electrons. The spins of the 100 keV electrons are rotated from the longitudinal into the transverse direction using the Wien filter. The two spectra are from two silicon surface-barrier detectors on opposite sides of the beam direction at a polar angle of 120°. The solid (black) and dashed (red) spectra have been taken with left and right circularly polarized laser light, respectively.

laser beam irradiating the cathode. Asymmetries from the integrals of these peaks have been determined for several target thicknesses. The extrapolation to zero thickness then yields the degree of polarization as the analyzing strength (Sherman function) is known for single scattering events only. The uncertainty of the determined polarization degree stems to a large extent from the lack of knowledge of the correct extrapolation of the asymmetries.

The test stand was operated also several times for first experiments in which the bremsstrahlung polarization orientation was measured depending on the orientation of the electron spin, see, *e.g.*, reference [8]. It can be shown that the orientation of the plane of linear polarization of the emitted bremsstrahlung at a given non-zero angle with respect to the beam axis is rotated depending on the degree of longitudinal (and transverse) polarization. Data analysis measuring the degree of linear polarization with a planar high-purity germanium 5x5 pixel detector and a lithium-drifted silicon 32x32 pixel detector are in progress. Beam currents in these experiments ranged from about 1 nA to a few μA with measurement times of several hours to about 1 day per setting.

3. Implementation at the S-DALINAC

Upon successful demonstration of the functionality of the source, it was dismantled at the teststand site and rebuilt at the normal conducting injection beam line of the S-DALINAC. To that end, the existing beam line between the thermionic unpolarized gun and the superconducting injector was removed. The inset in figure 1 shows a schematic drawing of the source at the new location. A second differential pumping stage separates now the thermionic gun from the UHV part of the beam line of the electron source. The preparation chamber was re-arranged, and the downstream differential pumping stage was rotated due to spatial restrictions. Also, the Mott polarimeter is installed in a vertical beam line following a 90° bending magnet due to spatial constraints.

Between the 270° alpha deflection magnet and the entrance into the superconducting injector linac, the beam line is used both for the polarized beam from the new source as well as the beam from the thermionic gun. While the polarized beam has a kinetic energy of 100 keV, the thermionic gun delivers 200 keV electrons, a slight reduction with respect to the former operation voltage of −250 kV. Hence, the beam line must accomodate both beam energies. Besides steering and focusing magnets, this is of particular importance for the radio-frequency (RF) installations.

Although a 3 GHz pulsed operation of the new source had already been shown to be possible from pulsing the diode laser system and utilizing a fast coaxial Faraday cup [12], a new chopper will ensure electron bunch lengths of at most 50 ps. The device consists of a copper cavity, and the design was adapted from the chopper used at MAMI [15]. The functionality of the device had already been demonstrated for the 100 keV beam at the polarized source teststand. To ensure good and efficient acceleration, this bunch length needs to be reduced to about 5 ps at the accelerator entrance. The compression is achieved using a two-stage harmonic prebuncher system, also adapted from the Mainz injector [16]. In addition to a 3-GHz copper cavity, a 6-GHz cavity provides for an increase of the linear regime used for the bunching process. All copper cavities are operated by new RF controls as generator-driven resonators. For the 6 GHz buncher, the standard RF system was modified to accomodate devices suitable for new frequency range. The magnet power supplies for deflecting, steering, and focusing the beam are driven by a new client-server control system via CAN bus.

The superconducting injector linac contained so far three niobium cavities: a five-cell capture cavity and two twenty-cell cavities for further acceleration. Each cavity was operated at a gradient of up to 5 MV/m in continuous-wave mode and had a fixed cell length of 5 cm, corresponding to a phase velocity of the electron bunches of $\beta = v/c = 1$. However, the injection velocity of the polarized beam with an energy of 100 keV only will be lower so that an additional two-cell capture cavity with cell length of 42.5 mm (corresponding to an average electron velocity of $\beta = 0.85$) has been re-installed to reduce phase slippage in the cavities. This capture cavity was first designed and operated in the 1990s [17] and was re-installed with the new source.

For polarization determination during the future operation of the S-DALINAC's polarized source, several polarimeters are being constructred and set up. Besides the 100 keV Mott polarimeter in the normal conducting part of the injector beam line (see (1) in figure 1), another Mott polarimeter has been installed at the experimental site behind the superconducting injector linac where up to 10 MeV electron beams are available (figure 1, (2)). The construction consists of two scintillators at a backward angles of $\pm 165°$. Monitoring of the degree of polarization at the injector experimental site will will realized using a Compton-transmission polarimeter described in reference [9]. Behind the main accelerator section, electron beam polarization will be determined by a Møller polarimeter (3) that is being designed for energies between 50 MeV and 130 MeV. Following a basic magnet design, the layout of the detector system is being prepared as is the connection to the accelerator beam line. The challenge with the Møller polarimeter is the low electron beam energy from which large scattering angles of the scattered and recoiling electrons result. At the tagger as well as the QClam spectrometer (points (4) in figure 1) additional Compton-transmission polarimeters are foreseen.

4. Laser system research and development

The new S-DALINAC polarized injector is driven by a laser system based on either a (pulsed) diode laser or a mode-locked Ti:Sapphire laser. Due to the little space available for laser research and development at the site of the source, the laser systems are located at a separate laboratory about 40 m away (cf. figure 1). The diode laser beam is transferred to the source using a polarization-conserving mono-mode fibre, whereas the intense short pulses of the Ti:Sapphire laser (repetition rate 75 MHz, the 40th sub-harmonic of the S-DALINAC's operation frequency) are guided in an evacuated tube. For the latter, an active stabilization with a pointing and centering system is under development. A test stabilization system has already achieved sufficient damping of external perturbations up to a frequency of about 500 Hz.

For the optimization and control of the laser systems, a spectrometer, a polarimeter, and – for the mode-locked operation of the Ti:Sapphire system – an autocorrelator have been developed or are being developed. For the future operation of the source, we are also investigating the possibility of using mode-locked diode laser systems at a repetition frequency of 3 GHz.

5. Outlook

A source of polarized electrons has been tested and installed at the superconducting Darmstadt electron linear accelerator S-DALINAC. We anticipate the commissioning of the new source early in 2011 following the next beam time and the commencement of experiments with polarized beams soon thereafter. Prior to the polarized beam commissioning, the present schedule foresees the operation of the new normal conducting injector beam line with unpolarized electrons from the 200 keV thermionic source.

Acknowledgments

This work builds on contributions by former students and co-workers, especially C Heßler and B Steiner, and on numerous discussions with H Genz, H-D Gräf, and A Richter. We are indebted to the technical staff at the Institut für Kernphysik at TU Darmstadt for their dedicated work. Funding by Deutsche Forschungsgemeinschaft through SFB 634 is gratefully acknowledged.

References

[1] Richter A 1996 *Proc. EPAC 1996* p 110

[2] Mohr P, Enders J, Hartmann T, Kaiser H, Schiesser D, Schmitt S, Volz S, Wissel F and Zilges A 1999 *Nucl. Instrum. Methods in Phys. Research* A **423** 480

[3] Savran D, Lindenberg K, Glorius J, Löher B, Müller S, Pietralla N, Schnorrenberger L, Simon V, Sonnabend K, Wälzlein C, Elvers M, Endres J, Hasper J and Zilges A 2010 *Nucl. Instrum. Methods in Phys. Research* A **613** 232

[4] Yevetska O, Watzlawik S, Ahrens J, Alkhazov G D, Chizhov V P, Maev E M, Neumann-Cosel P von, Orischin E M, Petrov G E, Porté J-M, Richter A, Sarantsev V V, Schrieder G and Smirenin Yu V 2010 *Nucl. Instrum. Methods in Phys. Research* A**618** 610

[5] Neumann-Cosel P von, Richter A, Schrieder G, Shevchenko A, Stiller A and Arenhövel H 2002 *Phys. Rev. Lett.* **88** 202304

[6] Ryezayeva N, Arenhövel A, Burda O, Byelikov A, Chernykh M, Enders J, Grießhammer H W, Kalmykov Y, Neumann-Cosel P von, Özel B, Poltoratska I, Pysmenetska I, Rangacharyulu C, Rathi S, Richter A, Schrieder G, Shevchenko A and Yevetska O 2008 *Phys. Rev. Lett.* **100** 172501

[7] Burda O, Botha N, Carter J, Fearick R W, Förtsch S V, Fransen C, Fujita H, Holt J D, Kuhar M, Lenhardt A, Neumann-Cosel P von, Neveling R, Pietralla N, Ponomarev V Yu, Richter A, Scholten O, Sideras-Haddad E, Smit F D and Wambach J 2007 *Phys. Rev. Lett.* **99** 202502

[8] Tashenov S, Bäck T, Barday R, Cederwall B, Enders J, Khaplanov A, Poltoratska Yu, Schaessburger K-U, Stöhlker Th and Surzhykov A 2010 *Phys. Rev. Lett.* submitted

[9] Barday R, Bangert P, Enders J, Göök A, Aulenbacher K, Nillius F, Surzhykov A, Jakubassa-Amundsen D and Yerokhin V A 2010 *J. Phys: Conf. Series*, these proceedings

[10] Enders J, Bahlo T, Bangert P, Barday R, Brunken M, Eckardt C, Eichhorn R, Espig M, Freudenberger M, Göök A, Ingenhaag C, Lindemann J, Neumann-Cosel P von, Oberstedt A, Oberstedt S, Poltoratska Y, Richter A, Roth M, Schneider F, Wagner M, Weber A and Zwicker B 2010 *Proc. SPIN Physics Symposium Jülich 2010, J. Phys: Conf. Series*, submitted

[11] Aulenbacher K, Nachtigall Ch, Andresen H G, Bermuth J, Dombo Th, Drescher P, Euteneuer H, Fischer H, Harrach D v, Hartmann P, Hoffmann J, Jennewein P, Kaiser K H, Köbis S, Kreidel H J, Langbein J, Petri M, Plützer S, Reichert E, Schemies M, Schöpe H-J, Steffens K-H, Steigerwald M, Trautner H and Weis Th 1997 *Nucl. Instrum. Meth. in Phys. Research* A **391** 498

[12] Heßler C 2008 *Doctoral Dissertation* (TU Darmstadt; Munich: Dr. Hut)

[13] Steiner G B 2008 *Doctoral Dissertation* (TU Darmstadt; Göttingen: Sierke)

[14] Franke S, Ackermann W, Weiland T, Enders J, Heßler and Poltoratska Y 2009 *Proc. PAC 2009* p 4513

[15] Braun H, Herminghaus H and Streun A 1988 *Proc. EPAC 1988* p 997

[16] Shvedunov V I, Ihm M O, Euteneuer H, Kaiser K H and Weis T 1996 *Proc. EPAC 1996* p 1556

[17] Schardt P 1995 *Doctoral Dissertation* (TH Darmstadt)

Journal of Physics: Conference Series **298** (2011) 012003 doi:10.1088/1742-6596/298/1/012003

Upgrade of the inverted source of polarized electrons at ELSA

D. Heiliger, W. Hillert and B. Neff

University of Bonn, Physics Institute, Nussallee 12, 53115 Bonn, Germany

E-mail: heiliger@physik.uni-bonn.de

Abstract. Since 2000 an inverted source of polarized electrons at the electron stretcher accelerator ELSA routinely provides a pulsed beam with a polarization degree of about 80 %. One micro-second long pulses with 100 nC charge are produced by irradiating a strained-layer superlattice photocathode with laser light from a flashlamp-pumped Ti:Sa laser. A rectangular pulse shape is achieved by operating the source in space charge limitation.

The proposed hadron physics program requires an intensity upgrade to 200 mA which can be achieved by enlarging the emission area or by improving the quantum efficiency (QE). The resulting changes of the beam parameters (like emittance and space charge) and of the optics of the transfer line were investigated in numerical simulations.

In order to enhance the source performance a new load lock system with crystal storage and atomic hydrogen cleaning will be installed in the near future.

1. Introduction

Since 2006 experiments on baryon spectroscopy are performed at the University of Bonn, requiring circularly polarized photons which are generated by bremsstrahlung of longitudinally polarized electrons [1]. The polarized electrons cannot be produced via self-polarization according to the Sokolov-Ternov mechanism [2] due to a considerably long polarization time. Thus, in Bonn polarized electrons are generated in a dedicated source [3] and are transported to the experiment while aiming at the highest possible conservation of polarization.

The actual set-up of the source of polarized electrons and its load lock system is shown in figure 1. The operating chamber is build as an inverted structure. The main parameters of the source are determined by the properties of the injector chain of the ELSA stretcher ring. A beam energy of 48 keV is required for the buncher section of the pulsed injector linac and leads to a strongly space charge dominated beam transport to the linear accelerator. A pulse length of 1 µs and a repetition rate of 50 Hz are determined by the booster synchrotron.

Polarized electrons are generated by irradiating a strained-layer superlattice photocathode with circularly polarized laser light from a flash lamp pumped pulsed Titanium Sapphire laser. The generated laser pulse with a pulse length of 10 µs shows a spiking behaviour and is chopped into a 1 µs long pulse [4].

For the acceleration in the linear accelerator a rectangular current pulse is desirable. To generate a rectangular current pulse even though the laser pulse is not rectangular the emitted current (by default 100 mA) is controlled by space charge limitation. In order to vary the beam

Figure 1. Picture of the 50 keV source at the University of Bonn.

intensity the perveance[1] can be adjusted by changing the distance between the anode and the cathode.

For future hadron physics experiments a significantly higher beam intensity of approximately 200 mA is required. Such intensities will have an impact on the beam dynamics and the optics of the transfer line to the linear accelerator has to be optimized. In this paper, the results of numerical simulations of the strongly space charged dominated high current beam transport at 50 keV will be presented. In addition, the optics and beam diagnostics of the transfer line, indispensable for the verification of the optimal settings of the focussing and steerer magnets, will be shown.

2. Intensity upgrade

Since the end of 2009 the Be-InGaAs/AlGaAs-photocathode used in the last years was replaced with a new GaAs/GaAsP-photocathode, which has an eight times higher quantum efficiency and a larger emission area. With this new photocathode the desired current of 200 mA can be generated for regular operation. In the following the transfer line used for the transportation of the generated beam from the operating chamber to the entrance of the linar accelerator will be detailed. Afterwards the numerical simulation of the beam transport will presented.

2.1. Transfer line

Figure 2 shows a schematic drawing of the transfer line. Due to the pulsed operation and the resulting small cw-current-equivalent the lifetime of the photocathode is mainly limited by the partial pressures of contaminating gases like water vapour and carbon dioxide and not by ion back-bombardment. In the linear accelerator the pressure, especially of water vapour, is in the range of 10^{-7} mbar, because the existing structure cannot be baked-out. In order to avoid a degradation of the ultra high vacuum in the operating chamber[2] the transfer line has to provide

[1] The perveance is the constant of proportionality between the emitted current and the applied voltage and is only dependent on the geometry.

[2] There is a difference in the partial pressures of the contaminating gases of 6 orders of magnitude between the linear accelerator and the operating chamber.

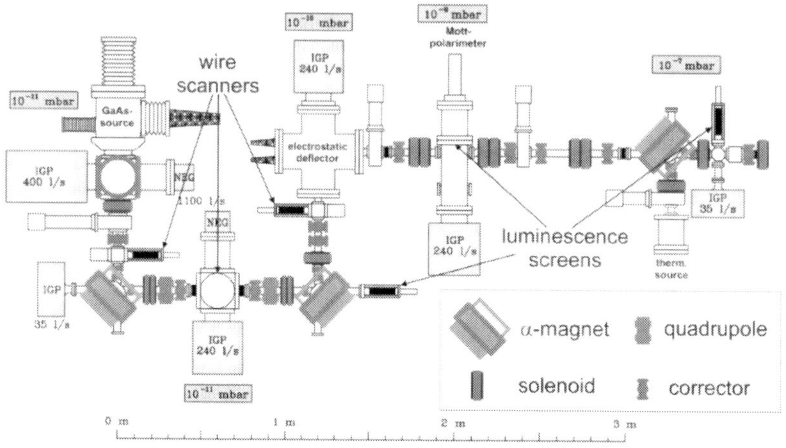

Figure 2. The transfer line between the operating chamber and the linear accelerator.

a 6 m long differential pumping section. Therefore the beam pipe has a diameter of 35 mm which is a compromise between a small aperture for differental pumping and a large aperture for the quasi lossless beam transport. In figure 2 the decrease in total pressure measured by dedicated vacuum gauges along the transfer line is indicated.

The folded beam line mainly consists of two α-magnets, an electrostatic deflector and solenoids and quadrupoles for focussing the beam. All in all there are 41 magnets in the transfer line and the beam is three times deflected by 90 degrees.

The set-up was chosen due to the following reasons:

- The electron beam can be easily separated from the laser beam.
- The height of the operating chamber of 1.5 m allows for an easy access.
- Backstreaming ions produced in the linear accelerator cannot hit the photocathode.
- A mirror symmetric magnetic optic (one symmetry plane is located between the α-magnets, a second in the Mott polarimeter) reduces the required beam diagnostics to three wire scanners and three luminescence screens.

α-magnets deflect the beam energy independent and act like a free drift, but provide a different effective drift length for each plane. Thus, a beam focus at the point of symmetry has to be achieved by using quadrupoles. Hence the cylinder symmetry[3] of the beam is disturbed which has to be taken into account for the numerical simulation of the beam transport (see chapter 2.2).

The electrostatic deflector rotates the longitudinal spin transverse to the momentum, which is necessary to conserve the polarization in the following circular accelerators. Behind the deflector mainly double solenoids are used for focussing the beam to achieve a vertically oriented spin at the injection into the booster synchrotron. The polarization is measured by means of a Mott polarimeter.

2.2. Beam transport

The transversal beam dynamics of a low-energy electron beam with a homogeneous, elliptical charge distribution in presence of external electromagnetic fields assuming a "laminar" flow are

[3] The emitting surface of the photocathode is a full circle resulting in a homogeneous, cylinder symmetric charge distribution of the beam when operating in space charge limitation.

described by the so called paraxial differential equations [5, 6]:

$$\frac{\mathrm{d}^2 x}{\mathrm{d}s^2} + [k_x(s) + S(s) + T(s)] \cdot x - \frac{\varepsilon^2}{x^3} - \frac{2K}{x+z} = 0, \qquad (1)$$

$$\frac{\mathrm{d}^2 z}{\mathrm{d}s^2} + [k_z(s) + S(s) + T(s)] \cdot z - \frac{\varepsilon^2}{z^3} - \frac{2K}{x+z} = 0. \qquad (2)$$

The linear term represents the restoring forces of quadrupoles ($k_{x,z}(s)$), solenoids ($S(s)$) and the electrostatic deflector ($T(s)$). The expansion of the beam due the emittance ε and the space charge[4] is included in the third and fourth term.

In order not to degrade the vacuum the beam must be transported quasi lossless to the linear accelerator. Due to the high intensity the beam transport is strongly space charged dominated. Thus, the focal length of the magnets are dependent on the actual beam current and have to be adjusted accordingly. In order to find out if a beam transport of 200 mA is feasible using the existing magnets in the transfer line, their focussing strengths were optimzed by iteratively feeding adjusted parameter sets into a program for the numerical solution of the differential equations. The optimization criteria were a minimal beam envelope along the whole transfer line and a beam focus in the symmetry planes.

Figure 3 shows the final results of the simulations and presents the beam envelope for optimal settings of the focussing strengths of the magnets for currents of 100 and 200 mA. Due to the geometry of the electrodes the beam is focused leading to a beam waist downstream and close by the anode. The waist position was chosen as the initial point of the simulation whereas its position varies for different currents and diameters of the emitting surface. The initial parameters like the position of the waist, the beam envelope at the waist and the emittance were taken from numerical simulations using the software EGUN [7].

In figure 3 the evolution of the beam envelope is presented by solid lines, above the abscissa for the horizontal and below the abscissa for the vertical beam plane. The shaded areas indicate the aperture of the transfer line. The origin of the diagram was set to the position of the first beam waist for 100 mA.

The beam envelope is always smaller than one third of the aperture in case of a beam current of 100 mA, which implies that a quasi lossless beam transport should be possible, assuming a well-defined beam shape (no halo) and the absence of strong disturbing fields. The operational experience with a current of 100 mA shows that an overall transfer efficiency close to 100 % could be obtained routinely and validates the simulation results.

The beam envelope for a current of 200 mA is larger than for 100 mA due to the higher space charge. Except near the alpha magnets, the envelope is always smaller than one half of the aperture, so that a quasi lossless transport appears to be feasible as well with 200 mA.

To verify the assumption of a cylinder symmetric, homogeneous and sharp-edged beam profile required by the paraxial differential equations and to adjust the focussing strength of the magnets based on the final results of the simualtion, dedicated beam diagnostics are needed. The available beam diagnostics are wire scanners and luminescence screens (see figure 2). Near the operation chamber only wire scanners are installed in order not to degrade the vacuum while scanning the beam profile. A wire scanner consists of two wires with a diameter of 50 µm mounted on a frame. The wire scanner is able to scan both planes of the beam by moving the wires through the beam and collecting the charge. The collected charge is converted into a voltage signal, amplified, integrated (every 20 ms for a duration of 1 µs) and digitalized. A very low background which can be reached by the use of a complete coaxial set-up of the scanners is of great importance due to the small currents of maximum 500 µA collected within the duration of scanning.

[4] K is called the generalized perveance and depends on the beam current and energy.

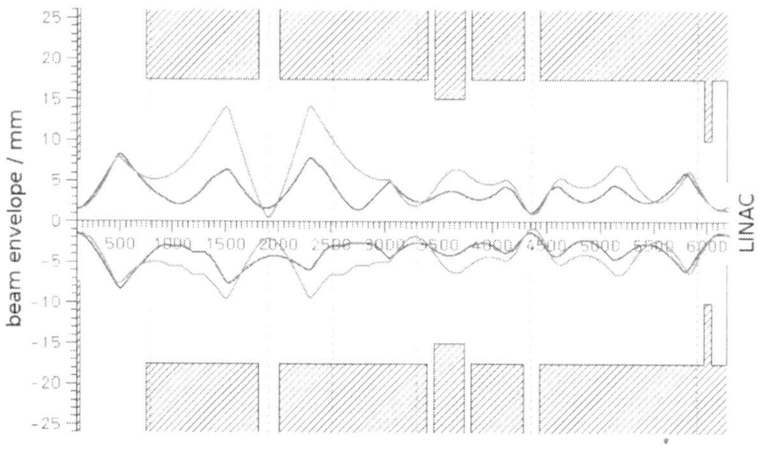

Figure 3. Final results of the numerical simulations for a beam current of 100 mA (blue lines, diameter of the photocathode Ø = 8 mm) and 200 mA (red lines, diameter of the photocathode Ø = 10 mm). Above the abscissa the horizontal and below the abscissa the vertical beam envelope is presented. The optimization criterium of a minumum beam envelope at the symmetry points is fullfilled for both currents.

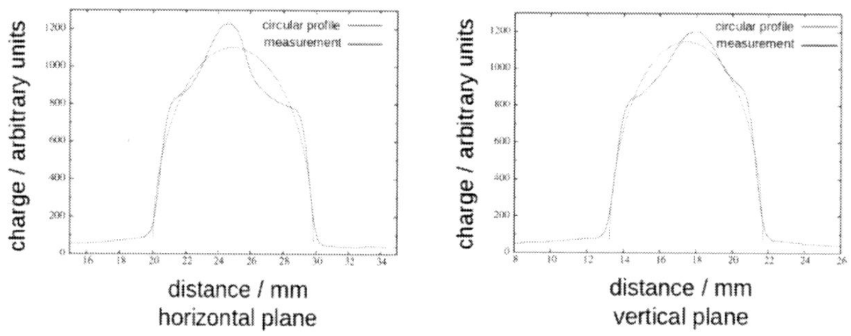

Figure 4. Beam profiles in both planes recorded with the first wire scanner in the transfer line.

The two beam profiles shown in figure 4 were recorded with the first wire scanner in the transfer line. Because the emitting surface of the photocathode is a full circle a homogeneous, cylinder symmetric charge distribution of the beam when operating in space charge limitation is expected before the beam has passed through the quadrupoles. The measurements are in good agreement with the expected well-defined beam profile (red curves) and legitimate the assumption for the paraxial differential equations. The slight charge redistribution visible in figure 4 is caused by inhomogeneous fringe fields of a permanent magnet of an ion getter pump.

3. New load lock system

In addition to the intensity upgrade an extreme high vacuum load lock system with an activation chamber, a storage chamber and a loading chamber (see figure 5) will be installed.

A new photocathode is brought into the loading chamber and is cleaned with atomic hydrogen while heating the photocathode to moderate temperatures ($< 400\,^\circ$C) for a better removal of surface oxidations. Afterwards it is transported to the storage chamber via an elevator. In the

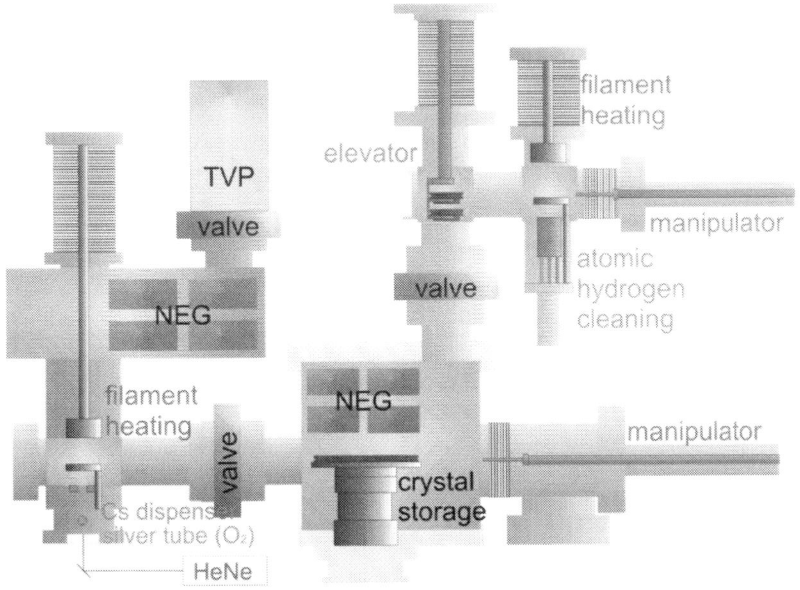

Figure 5. The new load lock system with an activation chamber, a storage chamber and a loading chamber with atomic hydrogen cleaning (from the left to the right).

storage chamber up to five photocathodes can be stored in extreme high vacuum comparable to the vacuum in the operating chamber. The activation is done with cesium and oxygen in the activation chamber, before the photocathode is transported into the gun for operation. The storage of up to five photocathodes will increase the reliability and uptime of the source, because the replacement of the photocathode in the operating chamber will take only a few hours compared to the 2 weeks required with the actual set-up. Futhermore the storing of different types of photocathodes with various diameters of the emitting surface will allow a quick change of the operation parameters like the emitted current and polarization degree. Additionally tests of the quantum efficiency, polarization degree and the success of atomic hydrogen cleaning of different types of photocathodes can be done during regular operation.

4. Conclusion

Since 2000 the source of polarized electrons provides very reliably a beam of 100 mA and a polarization degree of 80 %. The emitted current is controlled by space charge limitation. An intensity upgrade to a current of 200 mA is feasible with the actual set-up of the source and its transferline. Additionally an extrem high vacuum load-lock system with crystal storage and atomic hydrogen cleaning will enhance the source performance.

References

[1] Hillert W 2006 Eur. Phys. J. A **28** s01 139.
[2] Sokolov A and Ternov I 1964 Sov. Phys.-Doklady **8** 1203.
[3] Hillert W et al. 2001 AIP Conf. Proc. **570** 961.
[4] Eberhardt M et al. 2006 AIP Conf. Proc. **915** 1031.
[5] Buon J 1994 CERN Yellow Report CERN-**94-01** 89.
[6] Septier A, 1969 Academic Press.
[7] Herrmannsfeldt WB 1988 SLAC-PUB **331** UC28.

Polarized Electron Gun Development at the Brookhaven National Laboratory

J. Kewisch, X. Y. Chang, I. Ben-Zvi, V. Litvinenko, T. Rao, J. Skaritka, B. Sheehy, A. Pikin, W. Meng, E. Wang, Q. Wu, D. Pate, A. Burrill, T. Xin,

Brookhave National Laboratory, Upton, NY 11973, USA

D. Holmes

Advanced Energy Systems, Medford, New York 1176, USA

Jorg@BNL.GOV

Abstract. Development of two different polarized electron guns is ongoing at BNL. One aims at extremely high brightness at a moderate beam current. This design uses a superconducting RF gun and a test setup is built to show that a Gallium-Arsenide cathode with negative affinity has a sufficiently long quantum efficiency lifetime in such an environment. An electron injector using this technology may eliminate the need of the electron damping ring and a long transport line at the International Linear Collider. The other project aims at producing a high beam current with moderate emittance requirements, dubbed the "Gatling gun". In this DC gun, bunches are extracted from 20 separate cathodes and merged into a single beam using a rotating magnetic field. Such an electron gun could serve as an injector for the electron-ion collider eRHIC, which is planned at BNL. We will report on the status of these projects.

1. Introdution

Since the luminosity in a collider is proportional to the number of particles in each bunch and inversely proportional to the bunch area there are two ways to obtain high luminosity: One can increase the beam current or increase the brightness of the beams. When polarized electrons are required both these methods have their own set of problems, but they are ultimately limited by the rapid aging of the cathodes caused by ion back-bombardment. The number of ions which are created in the gun and hit the cathode, is proportional to the beam current and the vacuum pressure. A high beam current is therefore a direct cause of cathode aging. A high brightness requires a high accelerating voltage near the cathode, which can be obtained using an RF gun. However, normal conducting RF guns heat up during operation, which causes outgasing and a high vacuum pressure.

Two R&D projects at BNL are trying to address these problems: The "Polarized SRF Gun" project tries to eliminate the vacuum pressure problem of normal conducting guns by using a superconducting gun, which not only eliminates the outgasing, but also acts as a cryo-pump. The experiment is designed to show that the cathode in such a gun has a useful quantum efficiency lifetime and that the cesium layer does not contaminate the cavity.

The "Gatling gun" project is the construction of a DC gun with twenty separate cathodes. The beams from these cathodes are merged using a rotating magnetic field. The destruction caused by the back-scattered ions is therefore reduced for each cathode by a factor of twenty.

2. The polarized SRF gun

The setup consists of three parts: a cathode preparation chamber, which is used to clean and cesiate the cathode, the gun itself consisting of the beam line, Faraday cup and cryostat and a transporter which allows transferring the cathode from the preparation chamber to the gun without exposing it to air.

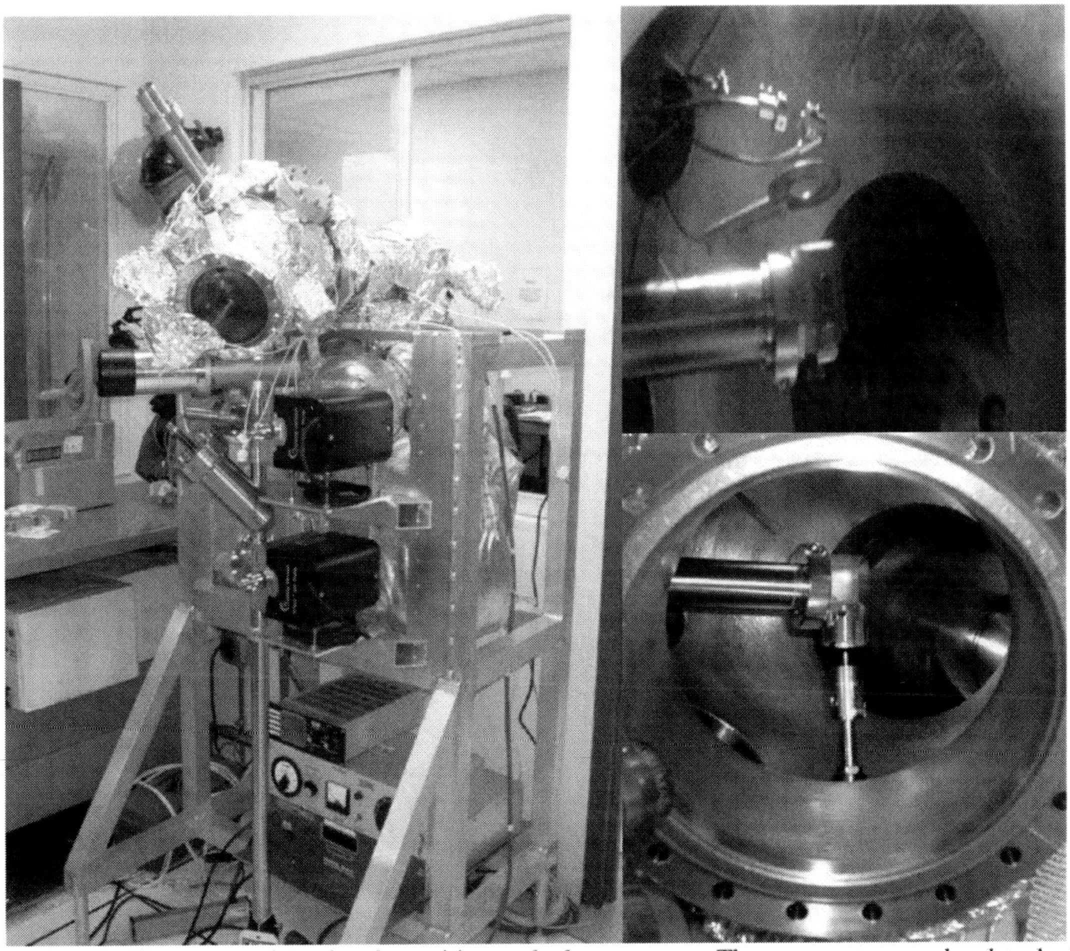

Figure 1: The preparation chamber with attached transporter. The transporter can be detached between the two valves for moving the cathode to the gun. The pictures on the right show the insertion of the cathode plug into the heater block and the location of the ring-shaped anode and holder of the cesium source.

Figure 1 shows the preparation chamber with attached transporter. During the cathode preparation the plug is inserted from below into a heater block in the preparation chamber (lower right picture). The cathode is heated to 580° Centigrade. The upper right picture shows the ring-shaped anode and the holder for the cesium source. A laser beam shines from above onto the cathode during preparation in order to measure the change of the quantum efficiency. O_2 is used instead of NF_3 in the cathode preparation, since oxygen does not require any special safety procedures.

Cathodes with a quantum efficiency of up to 10% have been prepared successfully. Figure 2 and 3 show the quantum efficiency as a function of time during the Yoyo procedure and during a preparation where cesium and oxygen are applied at the same time. With this procedure the final quantum efficiency is slightly better, which may be explained by the improved vacuum pressure. However, the preparation time is 30% shorter.

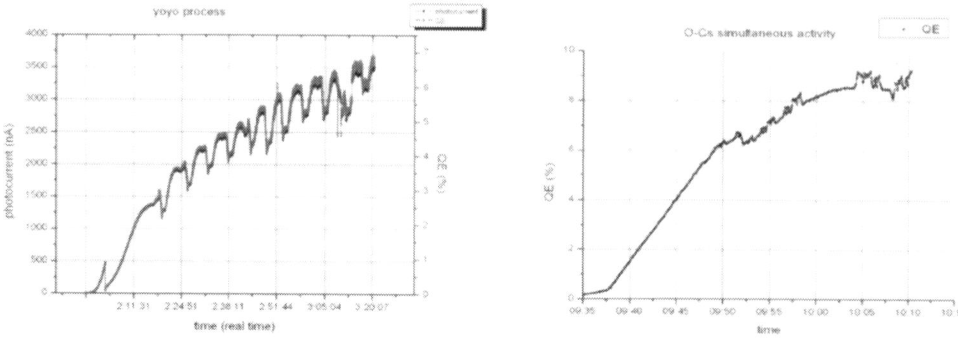

Figure 2: Quantum efficiency measured during Cesiation using the Yoyo procedure (left) and with deposition of cesium and oxygen at the same time (right).

The transporter consists of a movable rod, which is used to retract the cathode plug into the vertical pipe. Two ion pumps are used to keep the pressure in the transport chamber and the load lock chamber low. A NEG pump and a cold cathode gage with a Americium source where added to the original design. Although the manufacturer recommended not to exceed 250^{0} C for baking the pipe was baked at 400^{0} C without damage.

Figure 3: The gun assembly below (left) and above (right) of the top plate.

The gun assembly is shown in Figure 3. In the back wall of the gun is an adapter, which allows insertion of the cathode plug and a valve, which is opened after the transporter is attached and the volume between the valves is pumped down. After the insertion of the cathode into the gun the transporter is removed and replaced by a pipe which leads to an ion pump outside the cryostat. The pipe also serves as a pressure relief path in case the gun volume is filled with liquid Helium through a vacuum leak. NEG pumps above the gun improve the vacuum when the gun is warm. The electron beam exits through the top plate and is defected into a Faraday cup. The assembly of the beam line is completed. We are waiting for the completion of the radiation safety system and the Helium recovery system, which is expected by the end of 2010.

3. The "Gatling" gun

The Gatling gun project is in the designing and prototype constructing stage. The 200 kV DC gun will be designed to produce bunches of 3.5 nC with a bunch frequency of 14 MHz, resulting in a beam current of 50 mA. The final normalized emittance shall be not more than 20 mm mrad with a bunch length of 3 mm and an energy spread of 1% at 10 MeV. Figure 5 shows the layout of the accelerator. The 20 cathodes are arranged on a circle with 30 cm diameter. After acceleration and initial focusing, the beams are deflected to the center of the circle, where the beams are bent onto a common axis with time varying fields.

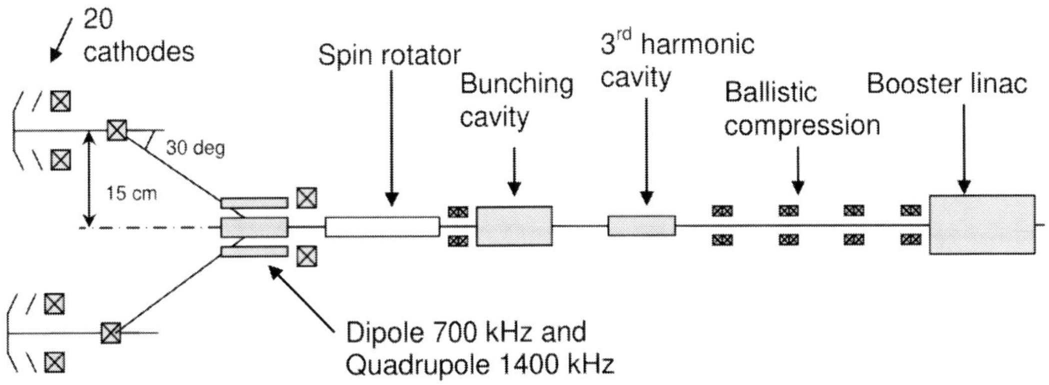

Figure 4: Layout of the Gatling gun accelerator.

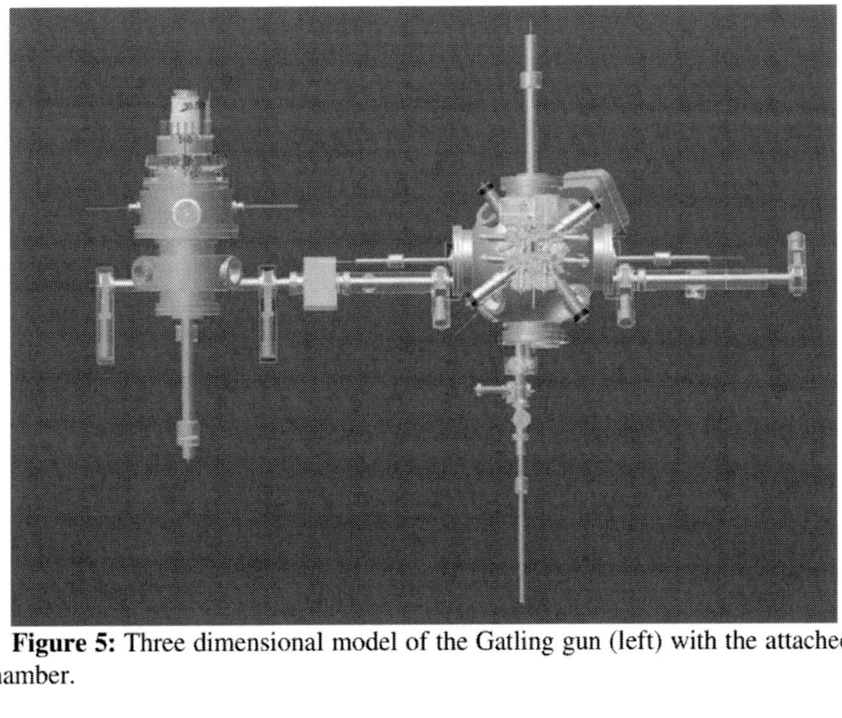

Figure 5: Three dimensional model of the Gatling gun (left) with the attached cathode preparation chamber.

Figure 6 show the three-dimensional model of the gun and preparation chamber. The beam will exit the gun towards the top. The green lines indicate the path of the laser beams. For the regeneration of the cathodes the cathode plugs are transferred from the circle onto a linear holder and moved through the connection pipe.

For spin dynamics reasons both deflections of the bunches have to be of the same type: either electric or magnetic. We use the magnetic deflection because the magnetic deflection is much more efficient than the electric deflection at our 200 keV of electron energy level. The combiner magnet has a ring shaped ferrite core with 20 separate coils which are powered with a frequency of 700 kHz and peak current of 124 Amperes AC current. With the correct phase relations this produces a rotating dipole field.

With only a dipole field the magnet also focuses, which makes the bunches elliptical. This increases the emittance of the combined beam. For compensation another 40 coils are added to create a rotating quadrupole, driven with 1400 kHz and 2.4 Amperes. Figure 6 shows a proto-type of the combiner magnet with 8 dipole coils and 16 quadrupole coils. Measurements show that the field quality of this magnet is sufficient not to increase the beam emittance.

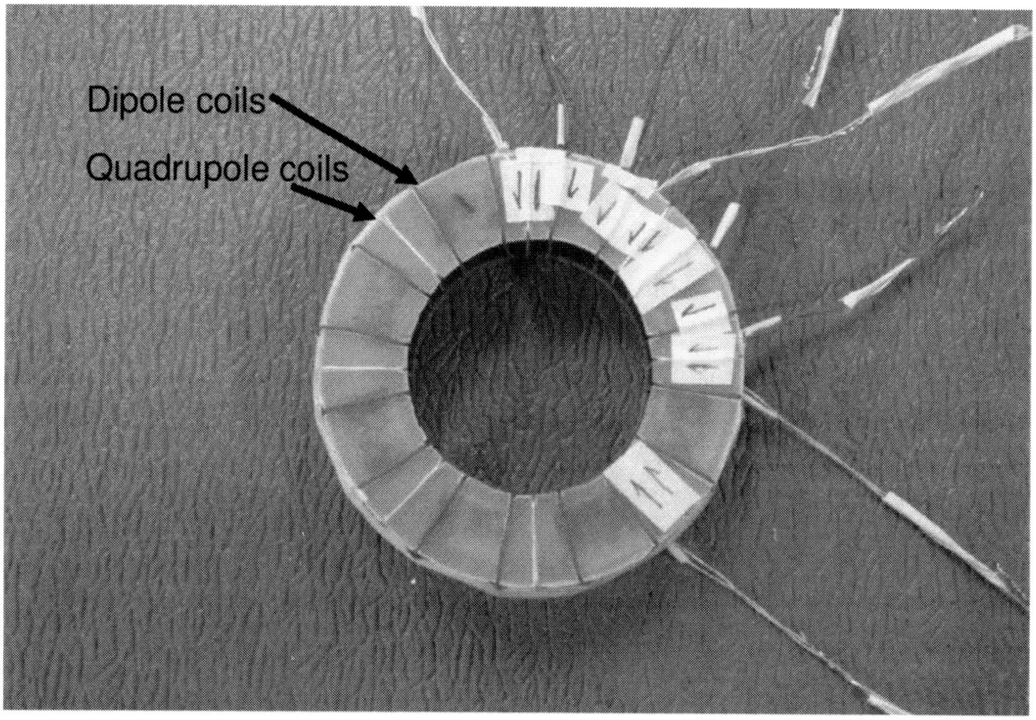

Figure 6: Prototype combiner magnet.

References

[1] Kewisch J, Ben-Zvi I, Rao T, Burrill A, Pate D, Grover R, Todd R, Bluem H, Holmes D and Schultheiss T 2007 *AIP Conference Proceedings* **980** 118

[2] Kewisch J, Ben-Zvi I, Rao T, Burrill A, Pate D, Wang E, Todd R, Bluem H, Holmes D and Schultheiss T 2008 *AIP Conference Proceedings* **1149** 1133

Journal of Physics: Conference Series **298** (2011) 012005 doi:10.1088/1742-6596/298/1/012005

DEVELOPMENT OF A 500-KV PHOTOCATHODE DC GUN FOR ERLS

N. Nishimori[1], R. Nagai[1], M. Yamamoto[2], Y. Honda[2], T. Miyajima[2], H. Iijima[3], M. Kuriki[3], M. Kuwahara[4], S. Okumi[4], T. Nakanishi[4], and R. Hajima[1]

[1] Japan Atomic Energy Agency (JAEA), Tokai, Naka, Ibaraki 319-1195, Japan
[2] High Energy Accelerator Research Organization (KEK), Tsukuba, Ibaraki 305-0801, Japan
[3] Hiroshima University, Higashihiroshima, Hiroshima 739-8530, Japan
[4] Nagoya University, Nagoya, Aichi 464-8601, Japan

E-mail: nishimori.nobuyuki@jaea.go.jp

Abstract. We have developed a 500-kV, 10-mA photocathode DC gun for energy recovery linac (ERL) light sources. A segmented ceramic insulator with guard rings is employed to improve robustness at high voltage operation, because this structure can prevent field emission electrons from directly striking the ceramic surface. We have recently succeeded in applying 500 kV on the ceramics for eight hours without any discharge. This high voltage testing was performed with a simple configuration without NEG pumps, cathode and anode electrodes to mainly study the field emission from a tube supporting the cathode electrode. The same high voltage testing with a full configuration necessary for beam generation was carried out up to 380 kV where some increase of radiation was observed. Up-to-date status of our gun development is presented in detail.

1. Introduction

Electron guns which can deliver a high brightness electron beam with emittance lower than 1 mm-mrad and current up to 100 mA are being developed for next generation Energy Recovery Linac (ERL) Light Sources (LS) worldwide [1, 2, 3]. A DC photoemission gun with a GaAs photocathode is one of the most promising candidates, since such a photoemission DC gun illuminated with 527 nm laser light has successfully provided 9.1 mA beam for the JLab 10 kW IR Upgrade FEL [3]. The gun high voltage equal to or greater than 500 kV is required to generate low emittance beam by reducing non-linear space charge effects in low energy region [4]. The accelerating field on the cathode surface should be as high as possible to suppress the space charge effects as well. This high brightness gun is anticipated to be used in a 5-GeV ERL based hard X-ray synchrotron light source project in Japan [1], an X-ray FEL oscillator as an attractive option of the 5-GeV light source [5], and an ERL based high-flux Compton gamma-ray as a new nondestructive assay method for ^{235}U, ^{239}Pu, and minor actinides in spent nuclear fuel assembly [6, 7].

We have developed a 500-kV DC gun for the Japanese ERL light sources [8]. One of technological challenges of high brightness DC guns is to apply DC high voltage on a ceramic insulator with a tube supporting cathode electrode, since field emission from the tube causes discharge or punchthrough on the ceramic surface [11]. In order to mitigate the field emission problem, a segmented insulator with rings which guard the ceramics from the field emission is employed. We have recently succeeded in applying 500 kV on the ceramics for eight hours without any discharge [9]. This high voltage testing was performed with a simple configuration without NEG pumps, cathode and anode electrodes to mainly study the field emission from the supporting tube. The next task is to apply the same high voltage on the ceramics with a full configuration necessary for beam generation. We have designed electrodes for the maximum surface electric field not to exceed 11 MV/m at 500 kV with the cathode/anode gap of 100 mm. NEG pumps with a pumping speed of 7200 l/s have been installed in the gun chamber. A photocathode preparation system was connected to the gun chamber. In the present paper, our current status of development of a photoemission DC gun is described.

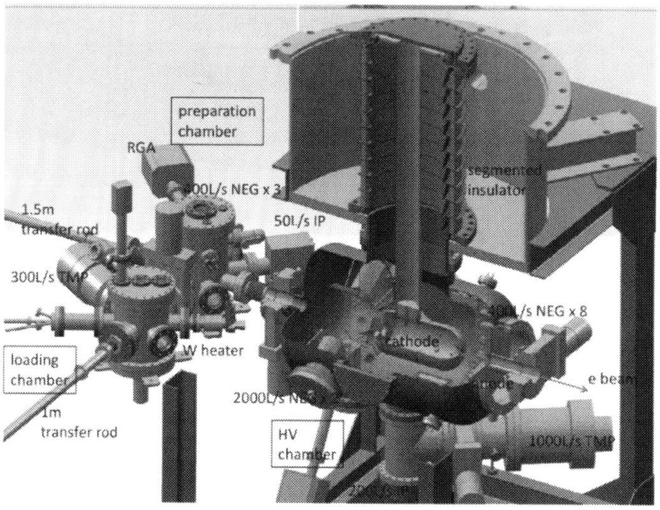

Figure 1. The 500-kV photocathode DC gun.

2. 500-kV DC photocathode gun

The 500-kV DC gun is shown in Fig. 1. A GaAs wafer on a molybdenum puck is installed in the loading chamber and heat cleaned by a tungsten heater. The puck is transferred by a magnetic coupling to the preparation chamber where cesium and oxygen are alternatively applied for negative electron affinity activation. The activated cathode is then transferred to the high voltage (HV) chamber and installed in the cathode electrode. The cathode electrode is connected to the high voltage terminal of the segmented ceramic insulator. The photoemission beam is accelerated by a static electric field applied between cathode and anode electrodes.

2.1. Gun electrodes

The bottom center of Fig. 2 shows cross section of the high voltage vacuum chamber around the cathode electrode. The cathode/anode gap is 100 mm to obtain an accelerating field as high as possible on the photocathode surface without high-voltage breakdown. A static electric field calculation (POISSON [10]) shows the maximum field on the cathode electrode and accelerating field on the photocathode center are 10.3 MV/m and 6.75 MV/m at 500 kV, respectively (see right of Fig. 2). The cathode/anode gap is surrounded by NEG pump units. Eight 400 l/s NEG pumps (SAES: CapaciTorr D400-2) were installed in the chamber. The unit of NEG pumps is shown in Fig. 3. The number of the pumps will be increased to twenty in the near future, which corresponds to 8000 l/s pumping speed for hydrogen. The NEG pumps are covered with mesh HV shields made of titanium wire with 1 mm in diameter.

Five ICF203 ports of the HV chamber, which are located behind the cathode electrode, are used to install 2000 l/s NEG pumps (SAES: CapaciTorr D2000). The distance between the cathode tail and NEG pumps along the beam line is 64.5 mm. The POISSON calculation shows the maximum electric field on the cathode electrode is 10.51 MV/m (see left of Fig. 2). Two 2000 l/s NEG pumps were already installed in the chamber. The number of the pumps will be increased to five in the near future. A 200 l/s ion pump (ULVAC:PST-200AU) is installed at the bottom of HV chamber (see Fig. 1) to pump noble gases and methane.

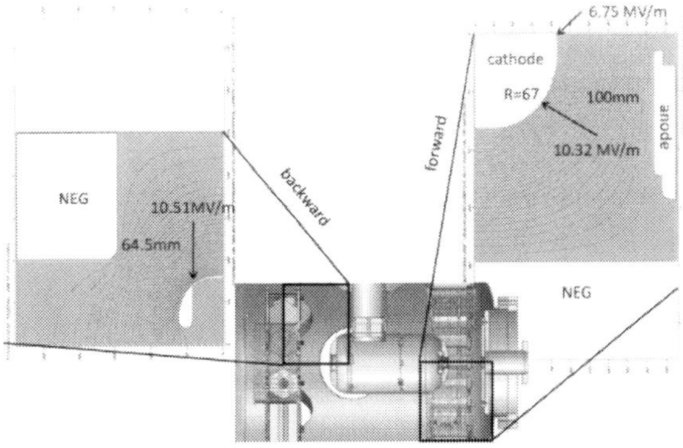

Figure 2. Static electric field calculation around cathode and anode electrodes.

Figure 3. A unit of five 400 l/s NEG pumps installed around the cathode/anode electrodes in the HV chamber.

The cathode electrode consists of three separate parts, head, body and tail. A photocathode is housed in the electrode head and the support tube is connected to the electrode body (see Fig. 1). All electrodes are made of titanium. We can easily switch to electrodes with different shapes and materials.

The ceramic insulator and the HV chamber were baked at 180°C for 50 hours after assembling the gun system. A 1000 l/s turbo molecular pump installed at the bottom of HV chamber is used to pump down the chamber during the baking. After the activation of NEG pumps, the base pressure of the HV chamber is 1.8×10^{-9} Pa measured with a BA gauge (ULVAC: AxTRAN).

2.2. High voltage conditioning

A high voltage conditioning was performed after the base pressure reaches below 2×10^{-9} Pa. We have to avoid discharge damage on the ceramics during the conditioning. Three techniques are employed to mitigate the damage. The first is a usual current limit register (100 MΩ) connected between the HV power supply and the insulator to limit abrupt current drawn from HV terminal in case of discharge. The next is a technique to clip average discharge current. The HV power supply needs to provide current running through its internal registers and the dividing registers of the segmented insulator. We set current limit at the current necessary for the high voltage set-point plus just 1 μA. If the average discharge current exceeds 1 μA, the HV power supply switches to constant current (CC) mode from constant voltage (CV) mode and the voltage drops quickly. This voltage drop helps to stop discharge. Then the voltage recovers gradually as the power supply charges its capacitor up. Finally the operational mode returns to CV mode. The last is an interlock system for vacuum and radiation. The high voltage power supply was interlocked with pressure above 1×10^{-6} Pa and radiation above 3 μSv/h. The BA gauge is used to monitor the pressure and a NaI radiation detector is placed 0.2 m downstream the anode flange to monitor the radiation increase due to discharge.

Figure 4 shows the applied voltage (top), radiation (middle) and pressure (bottom) as a function of time during the conditioning. The gun was conditioned up to 250 kV within two hours. Above 250 kV, it takes an hour to increase 4 or 5 kV up to 380 kV. The conditioning was halted at 380 kV, since radiation increase was observed above 370 kV. We have performed radiation survey around the HV chamber and found locally strong radiation at one of the five ICF203 ports which are used for installation of 2000 l/s NEG pumps (see Fig. 5). To further increase the high voltage, we need to open the high voltage chamber and fix the radiation problem. Before that, we plan to demonstrate beam generation at 300 kV.

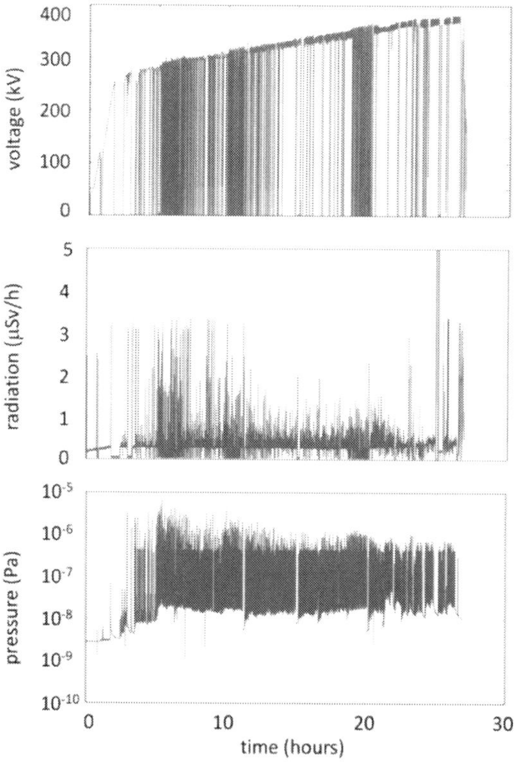

Figure 4. A high voltage conditioning result. The voltage (top), radiation (middle) and pressure (bottom) as a function of time.

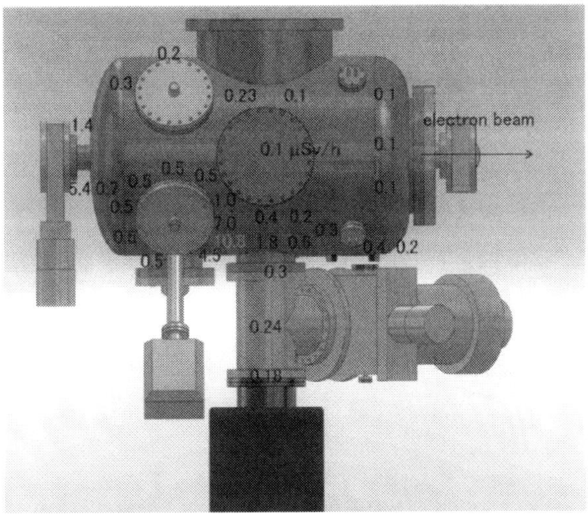

Figure 5. Radiation survey result for the right side of HV chamber when 370 kV is applied on the HV terminal. The figures represent radiation in units of μSv/h measured with a NaI detector.

3. Cathode loading and preparation system

A cathode preparation system was connected to the high voltage chamber. A 1.5 m magnetic coupling is used to transfer a puck between the preparation and high voltage chambers. The details of the system are described in Ref. [12]. A GaAs wafer installed into the loading chamber is heat cleaned to 550°C for five hours by a tungsten heater. The pressure rises to 3×10^{-5} Pa during the heat cleaning. The cleaned wafer is transferred to the preparation chamber, and then it is activated by alternative application of cesium and oxygen. The preparation chamber is equipped with three 400 l/s NEG pumps (SAES: CapaciTorr D400-2) and a 45 l/s ion pump (ULVAC: PST-050AU). The base pressure is 1.8×10^{-9} Pa.

The photocurrent is measured with a charge collector positively biased at 40 V, which is 1 cm in front of the photocathode, as a function of time under intermittent illumination of 20 μW laser light at wavelength of 532 nm. Typical quantum efficiency of 7 % at maximum is obtained. The static 1/e cathode life in the preparation chamber is 270 hours, as shown in Fig. 6.

4. Summary

We have developed a 500-kV photoemission DC gun. High voltage conditioning with full configuration for beam generation was carried out up to 380 kV, The conditioning was halted by radiation increase problem. We first plan to try beam generation at 300 kV and then to fix the radiation problem for further conditioning to 500 kV.

References

[1] Sakanaka S *et al.* 2010 Recent progress in the energy recovery linac project in Japan *Proc. of IPAC2010* 2338
[2] Dunham B M, Sinclair C K, Bazarov I V, Li Y, Liu X and Smolenski K W 2007 Performance of a very high voltage photoemission electron gun for a high brightness, high average current ERL injector *Proc. of PAC2007* 1224
[3] Hernandez-Garcia C, Siggins T, Benson S, Bullard D, Dylla H F, Jordan K, Murray C, Neil G R, Shinn M and Walker R 2005 A high average current DC GaAs photocathode gun for ERLs and FELs *Proc. of PAC2005* 1795
[4] Bazarov I V and Sinclair C K 2005 *Phys. Rev. STAB* **5** 034202

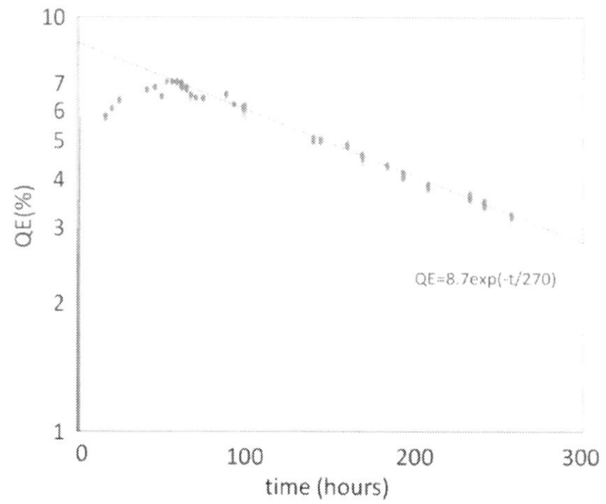

Figure 6. Static life of GaAs cathode measured in the preparation chamber. The decay time is 270 hours.

[5] Hajima R and Nishimori N 2009 Simulation of an X-ray FEL oscillator for the multi-GeV ERL in Japan *Proc. of FEL2009* 568

[6] Hayakawa T, Kikuzawa N, Hajima R, Shizuma T, Nishimori N, Fujiwara M and Seya M 2010 *NIMA* **621** 695

[7] Hajima R, Sawamura M, Nagai R, Nishimori N, Shizuma T, Hayakawa T and Seya M 2010 Design of a multi-turn ERL for hybrid K-edge densitometer *Proc. of FEL2010* 254

[8] Nishimori N, Nagai R, Iijima H, Honda Y, Muto T, Kuriki M, Yamamoto M, Okumi S, Nakanishi T and Hajima R 2009 *AIP Conf. Proc.* **1149** 1094

[9] Nagai R, Hajima R, Nishimori N, Muto T, Yamamoto M, Honda Y, Miyajima T, Iijima H, Kuriki M, Kuwahara M, Okumi S and Nakanishi T 2010 *Rev. Sci. Instrum.* **81** 033304

[10] Billen J H and Young L M 1996 *LA-UR* **96** 1834

[11] Nishimori N, Bazarov I, Dunham B, Grames J, Hernandez-Garcia C, Jones L, Militsyn B, Poelker M, Surles-Law K and Yamamoto M 2009 ERL09 WG1 summary: DC gun technological challenges *Proc. of ERL09* 6

[12] Nishimori N, Nagai R, Iijima H, Hajima R, Yamamoto M, Muto T, Honda Y, Miyajima T, Kuriki M, Kuwahara M, Okumi S and Nakanishi T 2009 Development of a 500-kV photo-cathode DC gun for the ERL light sources in Japan *Proc. of FEL2009* 277

Development of high brightness, high repetition rate photoelectron injectors at STFC Daresbury Laboratory

B L Militsyn[1], I Burrows[1], R J Cash[1], N Chanlek [1,3], B D Fell[1], L B Jones[1], S N Kosolobov[2], J W McKenzie[1], K J Middleman[1], H E Scheibler[2] and A S Terekhov[2]

[1]STFC Daresbury Laboratory, Warrington, WA4 4AD, UK
[2]Institute of Semiconductor Physics SB RAS, Novosibirsk, 630090, Russia
[3]University of Manchester, Manchester, M13 9PL, UK

E-mail : boris.militsyn@stfc.ac.uk

Abstract. Accelerator drivers for Energy Recovery Linac (ERL) and Free-Electron Laser (FEL) based light sources demand electron injectors which deliver high brightness bunches on the several hundreds of picocoulomb scale, at repetition rates between 1 MHz and 1 GHz (or higher), corresponding to an average current between 0.1 and 100 mA. Simultaneous satisfaction of these injector requirements is considerably beyond the current state-of-the-art. Daresbury Laboratory is concentrating efforts on the development of high average current III-V and $X_nY_{3-n}Sb$ photocathode-based DC and SRF photocathode guns for ERL applications. The ultimate goal of this research is their integration with the ALICE ERL.

1. Introduction

Low emittance high repetition rate electron beams are desired for several future Linac, Energy Recovery Linac (ERL) and FEL-based light source projects, including the UK's New Light Source (NLS) [1]. Currently operational linac-based FELs such as LCLS [2] and FLASH [3] use 1½ cell pulsed normal conducting RF guns operating at the frequency of the main linac. The LCLS S-band gun is powered by a short 2 μs RF pulse which permits operation in single pulse or short train mode. The LCLS injector has experimentally demonstrated beams with emittance values down to 0.3 mm mrad; however, its duty factor is limited by repetition rate due to cavity heating. For a normal conducting RF gun, the 1 ms duration of the RF pulse at FLASH is unusually high, though this permits gun operation in the long train pulse mode and at a pulse frequency in the train of 1 MHz which approaches a duty factor of 1%. Electron drivers of new generation soft X-ray FELs like UK's NLS, Wisconsin FEL and Berkeley FEL projects require operation at a repetition rate of 1 MHz or higher, delivering bunches with a charge of up to 1 nC. Injectors for those machines cannot be realised on the basis of traditional L- and S- band normal-conductive technology. This duty factor can be increased by using either DC or relatively-low frequency NCRF gun technology. The development of SRF technology raises the possibility of building CW electron guns.

The demands of new ERL projects are even higher than those examples given. Typical repetition rates vary around the linac frequencies with a typical bunch charge of 77 pC to deliver an average current of 100 mA. These current values are stated in the design of the Cornell University ERL [4], the BerlinPro machine [5] and others. In the framework of our ongoing project, Daresbury Laboratory is concentrating on the development of a macropulse photoinjector for the ALICE ERL, operating at a repetition rate of 81.25 MHz with a bunch charge of 80 pC which equates to 6.5 mA train current, and a CW photoinjector for the NLS project operating at 1 MHz with a bunch charge of 200 pC equating to an average current of 0.2 mA. An upgrade of ALICE to support CW operation demands average current of 1.6 mA. In Table 1, we summarise demands on the high brightness high repetition rate injectors for the accelerators currently under development in the world.

Table 1. Beam parameters required for high repetition rate injectors.

Beam Parameter	ERL (DL, Cornell,BerlinPro)	X-FELs (X-FEL,NLS)
Beam energy (MeV)	1-10	1-10
Maximum bunch charge (nC)	0.1	0.1-1
Bunch transverse emittance (mm mrad)	1.0	0.5-1
Maximum bunch repetition rate(MHz)	1300	1
Average current (mA)	10-100	1
Peak current of injector (kA)	-	0.1
Bunch duration (ps)	20	20

2. Electron emitters

The most critical component of high average current injectors is the photocathode. The minimum emittance which may be obtained from a photocathode with an internal energy E_i is defined by:

$$\varepsilon_{n,rms} = \sqrt{\overline{x^2}\,\overline{x'^2} - \overline{x \cdot x'}^2} = \sigma_\perp \sqrt{\frac{2E_i}{3mc^2}}$$

where $\sigma_\perp = \sqrt{q/\pi\varepsilon_0 E_c}/2$ is the minimal rms transverse beam size limited by the 2D charge density of a pancake beam. In Table 2 we summarise the ultimate emittances which may be obtained from a gun built with different technologies utilising photocathodes with $E_i = 1 \quad eV$, this being a typical value for many modern photocathodes operating in pulsed mode.

Table 2. Ultimate achievable emittance for different gun technologies.

Technology	Cathode field strength [MV/m]	Achievable emittance at a bunch charge [mm mrad]		
		0.01 nC	0.1 nC	1.0 nC
DC gun	10	0.11	0.34	1.08
VHF gun	20	0.08	0.24	0.77
L-band gun	50	0.05	0.15	0.48
S-band gun	100	0.03	0.11	0.34

The ALICE ERL at Daresbury operates in a pulse-train mode with a train current of 6.5 mA generated by a GaAs photocathode. Significant effort at the Laboratory is focused on the development of medium and high current photocathodes suitable for operation in the 1-10 mA current range.

2.1. III-V photocathode development

Historically, III-V family photocathodes such as GaAs, GaAsP, InGaAsP were used mainly in DC guns for the production of polarised electrons. As grown, these materials have a positive electron affinity (PEA), which is 4.8 eV for GaAs. In order to make GaAs photocathodes able to emit electrons when illuminated by 532 nm light, the electron affinity should be reduced to less than 1 eV - or even brought to a negative value. This activation process basically comprises deposition onto an atomically-clean photocathode surface of thin layers of Cs and an oxidant, typically O_2 or NF_3. Before the activation, the surface of the photocathode is chemically etched and heat cleaned in order to remove As and Ga oxides.

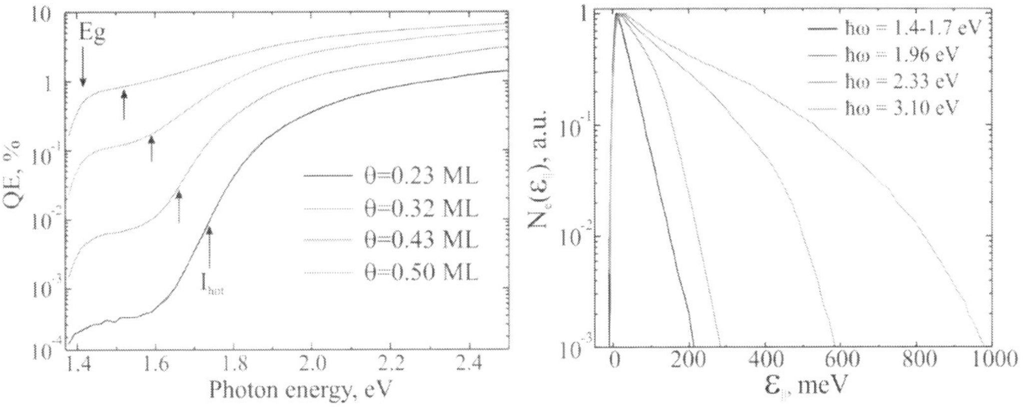

Figure 1. QE spectra of p-GaAs(Cs) –photocathode for different Cs coverage (left), and longitudinal energy distribution curves at different photon wavelengths for a PEA GaAs photocathode (right).

Recent measurements of the QE spectra indicate that GaAs activated to PEA with a *QE* of a few per cent and illuminated with green light is capable of picosecond level response times. This is enough to deliver bunches with a charge of several dozen pC. Figure 1 shows *QE* spectra for PEA GaAs photocathodes activated with Cs only to different levels of Cs coverage. The position of low energy threshold corresponds to the energy gap E_g for GaAs, and the position of I_{hot} corresponds to the vacuum level. The energy difference between these two thresholds is equal to the effective electron affinity. Figure 1 also shows the measured longitudinal energy distribution of electrons emitted from a PEA GaAs photocathode, demonstrating that when illuminated with green light, the energy spread is relatively large.

GaAs photocathodes place extremely high demands on operational vacuum conditions as they are extremely sensitive to the presence of oxidants in the residual atmosphere. For example the 1/e lifetime of GaAs does not exceed $2{\cdot}10^{-8}$ mbar·s of oxygen exposition [6]. The pressure in a typical GaAs gun is of the order of 10^{-11} mbar. Low operational lifetime is also an issue because the dominant GaAs degradation mode is bombardment of the cathode surface by back-streaming ions.

2.2. *Alkali photocathode development*

A good alternative to GaAs for operation at high and medium average currents may be antimonide-

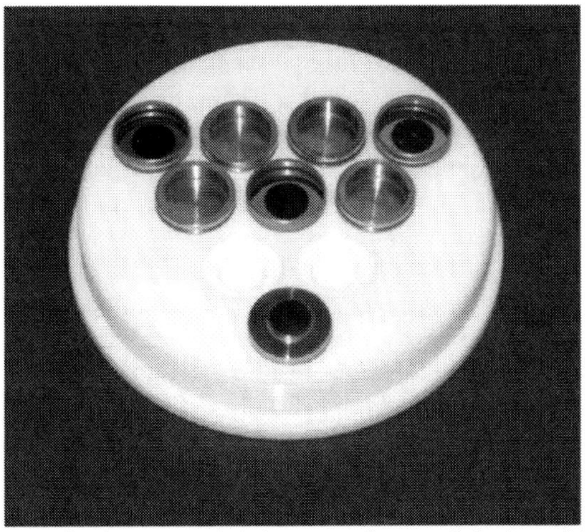

Figure 2. GaAs photocathodes on molybdenum substrated designed for the ALICE gun upgrade.

based photocathodes. Alkali photocathodes (AP) operate in visible and near-infrared light and demonstrate quantum efficiency of up to 80 mA/W or 18% at 532 nm wavelength. As most of these photocathodes exhibit Positive Electron Affinity (PEA), they are able to maintain fast response time (in the ps range on the metal substrate) even at high quantum efficiency levels in contrast to Negative Electron Affinity (NEA) photocathodes like GaAs. The dark lifetime of these photocathodes is three orders of magnitude better than GaAs [7], and the vacuum demands are far less stringent. However, their operational lifetime, or lifetime under ion back-bombardment, is unknown, and this has to be investigated. AP photocathodes have been used in photomultiplier tube (PMT) industry for many years, and this technology is well developed. AP have been used in an accelerator injector only once: in 1994 the Boeing project used an AP which set the world record of 32 mA current in an RF CW gun [8]. Presently, AP are considered as the photocathode of choice where a current exceeding 1 mA is required, such as ERLs (BerlinPro, BNL) and high repetition rate FELs (LBNL, Wisconsin). Cornell University, JLab and other organization are also interested in the development of the AP technology. Daresbury Laboratory plans to begin work on AP in the next financial year.

3. DC photocathode guns based photoinjectors

High voltage DC electron guns with GaAs photocathodes are considered as the gun of choice by a number of laboratories worldwide as injectors for ERL-based free electron lasers due to their potential to deliver beams of high average current (up to 100 mA in CW mode) with a relatively low normalized emittance of a few mm mrads. Since the minimum emittance, ε_{min}, of the delivered electron beam is related to the electric field strength on the cathode surface as $\sim 1/\sqrt{E_c}$ there is a strong motivation for increasing gun operational voltages. At TJNAF [9], Daresbury Laboratory [10] and JAEA/KEK [11], the gun power supplies are rated to 500 kV whilst the Cornell University injector [4] is rated to 750 kV.

3.1. *ALICE Gun*

The ALICE electron gun at Daresbury Laboratory is a modified version of the TJNAF design developed for Infra-Red FEL, and is described in detail elsewhere [12]. The main modification to the original TJNAF gun design is the use of a single large ceramic with bulk-doped controlled resistivity as the high voltage insulator. Whilst initially successful, with routine conditioning to 450 kV (up to a maximum of 485 kV), the long-term reliability of the brazing joints under load due to thermal cycling during baking has been poor. A collaboration between Daresbury Laboratory, TJNAF and Cornell University has resulted in design of an insulator with a modified taper at the brazed joint. This insulator has been manufactured, and is planned for installation as part of the gun upgrade.

3.2. *ALICE Gun Upgrade*

An upgrade to the ALICE gun is currently underway, and is schedule for installation in 2011. This will involve the installation of a new high voltage insulator, the development of new photocathodes, the addition of an external "load-lock" photocathode preparation facility, and a side-loading transport mechanism for inserting/extracting photocathodes to/from the gun. An extended gun beamline incorporating a suite of diagnostics useful both for ALICE operations and the testing of different photocathodes is also under consideration and is described in [13]. A detailed description of the gun upgrade and photocathode preparation facility may be found in [14-15].

ALICE currently uses GaAs photocathodes with an exposed diameter of 25 mm. The new cathode assembly of the gun, however, has been designed to accommodate photocathodes with an emission surface diameter of 10 mm. This is because the laser only illuminates an area ~ 4 mm in diameter and a smaller active area could help reduce the beam halo. New photocathodes designed for ALICE are shown in Figure 2. The preparation facility allows a variety of III-V photocathodes to be activated and tested in the ALICE photoinjector, these having various active layer composition, thickness, and electron affinity. Beam dynamics simulations have shown that the ultimate beam emittance which may be obtained from the gun is less than 1.0 mm mrad [15].

4. SRF gun based photoinjectors

The next project developed within the Daresbury photoinjector program framework is the design of a high repetition rate high brightness photoinjector for ongoing the New Light Source project. The machine requires 200 pC bunches at a repetition rate of 1 MHz, equating to an average current of 0.2 mA. The emittance of this beam should be as low as 0.5 mm mrad. At 1 MHz CW, existing UV laser systems are not able to drive metallic photocathodes, so Cs_2Te alkali photocathodes are considered instead. A 3½ cell SRF gun operating with a Cs_2Te photocathode is under commissioning at FZD [16] with a goal to deliver 1 mA average current.

To reduce complexity of the injector, we focus on a single cavity, 1½ Tesla cell design [17]. Recently a prototype of such a cavity was vertically tested to a peak electric field of 40 MV/m [18], and a gun with a field of 50 MV/m is potentially feasible. However, as this is pushing the state-of-the-art, beam dynamics simulations have been carried out at both 40 and 50 MV/m peak fields.

Since the high acceleration field in the SRF gun prevents bunch expansion and keeps bunch length close to that of the initial laser pulse, a separate buncher is not required. ASTRA [19] simulations have been carried out with a laser pulse having a 1 mm diameter flat-top transverse profile and a 30 ps flat-top temporal profile, with rise and fall times of 2 ps. Initial thermal energy of 0.7 eV was included, based on the use of a Cs_2Te photocathode. After the gun cavity, there is an emittance compensation solenoid which may be superconducting so that it can be contained within the gun cryostat. An alternative is to have a second cavity operating in a TE mode which provides a field profile similar to that of a solenoid. This would require additional cavity design and a dedicated RF power supply locked to the gun RF, due to the different frequency to the TM accelerating mode.

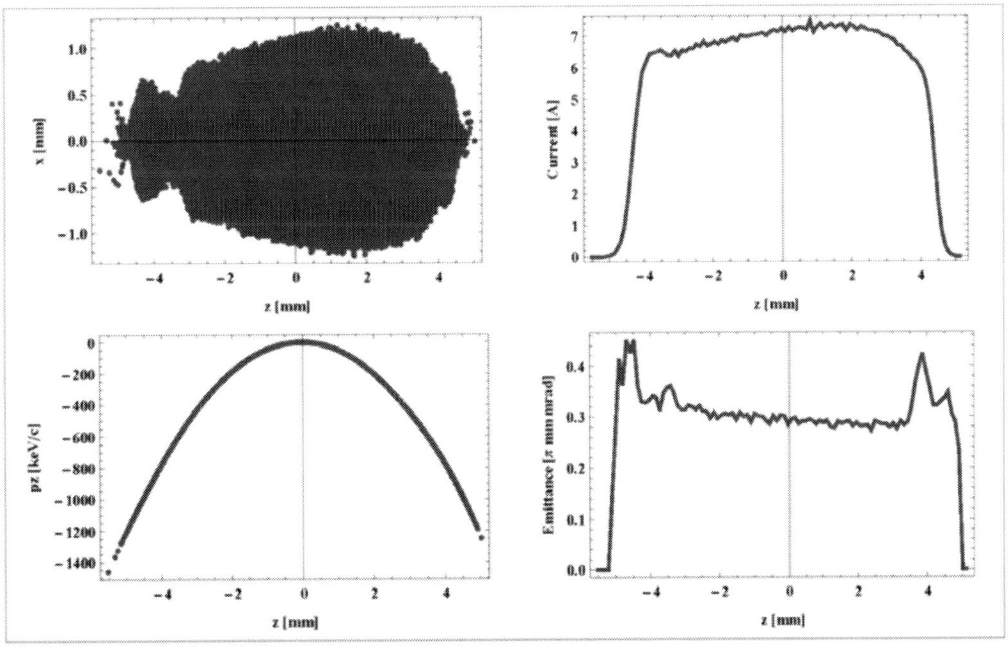

Figure 3. Bunch properties for the injector at the exit of the first linac module, with gun peak fields of 50 MV/m.

At a peak field of 50 MV/m, the gun accelerates the electron beam to 4.5 MeV. Operation at 40 MV/m reduces this energy to 3.6 MeV. The gun is operated at a phase of +3° from crest for 50 MV/m and +1° for 40 MV/m. These phases were chosen for optimal emittance. Following the solenoid, the beam is then injected into the first module of the main linac where it is further accelerated to 130 MeV. The linac module contains eight 9-cell Tesla cavities at a frequency of 1.3 GHz. These are operated at an average accelerating gradient of 15 MV/m. The first cavity is operated at a phase of -40°. The remaining cavities are then operated at +4.5° in order to compensate the energy chirp. Figure 3 shows the final distribution of the 200 pC bunch at the exit of the first linac module for ASTRA simulations at 50 MV/m. The flat current profile has duration of 35 ps and the slice emittance around 0.3 mm mrad throughout the bunch.

5. Summary

Ongoing and perspective projects stimulate development of high brightness high repetition rate injector at STFC Daresbury Laboratory. The GaAs-based 350 kV DC gun for ALICE ERL has been operational since August 2006. Gun upgrade stimulated development of a photocathode preparation facility has been recently successfully commissioned. The photocathodes have been activated with a maximum QE of 15% measured at 635 nm. Work on NLS project motivates SRF technology in the Laboratory and on the basis of this technology development of a high brightness high repetition rate gun.

6. References

[1] Aeppli G, Alleman R, Almond A, Angal-Kalinin D, Ashfold M, Averbukh V, Bagshow C, Barahona M, Barker P, Bartolini R *et al* 2010 *NLS Project: Science Case & Outline Facility Design* ed J Marangos, R Walker *et al* http://www.newlightsource.org
[2] Emma P, 2009 *Proc. of PAC 2009(Vancouver, Canada)* pp 3115-9
[3] Honkavaara K, Faatz B and Schreiber S 2008 *Proc. of FEL 2008 (Gyeongju, Korea)* pp 533-6
[4] Dunham B M, Sinclair C K, Bazarov I V, Li Y, Liu X and Smolenski K W 2007 *Proc. of PAC 2007(Albuquerque, NM)* pp 1224-6
[5] Abo-Bakr M, Anders W, Kamps T, Knoblach J, Kuske B, Kugler O, Matveenko A, Meseck A, Neumann A and Quast T 2009 *Proc. of SRF 2009(Berlin, Germany)* pp 223-7
[6] Pastuszka S, Terekhov A S and Wolf A 1996 Stable to unstable transition in the (Cs, O) activation layer on GaAs (100) surfaces with negative electron affinity in extremely high vacuum *Appl. Surf. Sci.* **99** 361
[7] Michelato P, Di Bona A, Pagani C, Sertore D and Valery S 1996 R & D activity on high QE alkali photocathodes for RF gun *Proc. of PAC 1995(Dallas, TX)* pp 1049-51
[8] Dowell D H, Bethel S Z and Friddell K D 1996 Results from the average power laser experiment photocathode injector test *Nucl. Instr. and Meth.* A **356** 167-76
[9] Siggins T, Sinclair C, Bohn C, Bullard D, Douglas D, Grippo A, Gubeli J, Kraft G A and Yunn B 2001 Performance of a DC GaAs photocathode gun for the Jefferson Lab FEL *Nucl. Instr. and Meth.* A **475** 549-53
[10] Saveliev Y M, Holder D J and Muratori B D 2008 *Proc. of EPAC08(Genoa, Italy)* pp 208-10
[11] Nishimori N, Nagai R, Iijima H, Hajima R, Yamamoto M, Muto T, Honda Y, Miyajima T, Kuriki M, Kuwahara M *et al* 2009 *Proc. of FEL 2009 (Liverpool, UK)* pp 277-80
[12] Gerth C and Hannon F E 2004 *Proc. of EPAC 2004 (Lucerne, Switzerland)* pp 1-3
[13] McKenzie J W, Muratori B D and Saveliev Y M 2009 Extended ALICE Injector *Proc. of PAC 2009 (Vancouver, Canada)* pp 4363-5
[14] Militsyn B L, Burrows I, Cash R J, Fell B D, Jones L B, McKenzie J W, Middleman K J, Kosolobov S N, Scheibler H E and Terekhov A S 2008 Design of an upgrade to the ALICE photocathode electron gun *Proc. of EPAC08 (Genoa, Italy)* pp 235-7
[15] Militsyn B L, Burrows I, Cash R J, Fell B D, Jones L B, McKenzie J W, Middleman K J, Scheibler H E and Terekhov A S 2010 High voltage DC Photoinjctor Development at Daresbury Laboratory *ICFA Beam Dynamics Newsletter* **51** ed S Chattopadhyay pp 161-70
[16] Teichert J, Arnold A, Büchner A, Büttig H, Janssen J, Justus M, Lehnert U, Michel P, Murcek P, Schurig R *et al* 2008 *Proc. of EPAC08 (Genoa, Italy)* pp 2755-7
[17] McKenzie J W and Militsyn B L 2010 *Proc. of IPAC'10 (Kyoto, Japan)* pp 1746-8
[18] Neumann A, Anders W, Dirsat M, Frahm A, Jankowiak A, Kamps T, Konoblach J, Kugeler O, Quast T, Rudolph J *et al* 2010 *Proc. of LINAC 2010 (Tsukuba, Japan)*
[19] Flöttmann K ASTRA homepage http://www.desy.de/~mpyflo

The ALICE Energy Recovery Linac – Project overview and injector performance

LB Jones, JW McKenzie, KJ Middleman, BL Militsyn, YM Saveliev and SL Smith

STFC Daresbury Laboratory, The Cockcroft Institute, Daresbury Science and Innovation Campus, Warrington, Cheshire, WA4 4AD, United Kingdom

E-mail: lee.jones@stfc.ac.uk

Abstract. The ALICE accelerator (**A**ccelerators and **L**asers **I**n **C**ombined **E**xperiments) at Daresbury Laboratory is a 35 MeV ERL. The electron beam drives an infra-red free-electron laser (FEL) and THz light sources, but can also be used to generate X-rays through Compton back-scattering (CBS). ALICE also acts as the injector for the EMMA NS-FFAG[1] machine. This paper will outline the project status and milestones achieved thus far, focussing on the status and performance of the photoinjector gun and the injection line.

1. Introduction

The ALICE ERL has a broad role as an accelerator physics and technology test facility, leading to the development of a full-scale ERL-based light source in the United Kingdom. Since the status of this project was last reported to this workshop [1], a number of significant milestones have been achieved such as the first beam acceleration in an SRF linac (23rd October 2008), followed quickly by full energy recovery (12th December 2008), generation and exploitation of THz radiation, and the generation of CBS X-rays [2].

2. ALICE Accelerator systems overview

ALICE uses a DC photoinjector gun which is a variant of the gun developed for the IR-FEL at the TJNAF [3]. It is designed to operate at 350 kV with a reflection-mode GaAs photocathode, activated *in-situ* by means of a caesium-oxidant '*Yo-Yo*' procedure. However, due to repeated vacuum failures of the single-piece bulk-doped ceramic with which the gun was designed, the use of slightly smaller 2-piece ceramic currently limits the operating voltage to 230 kV.

The photoinjector drive laser beam is provided by a HighQ IC10000 Picotrain system, which uses diode-pumped neodymium yttrium vanadate (Nd:YVO_4) to deliver >10 W at 1064 nm wavelength, which is later frequency-doubled to 532 nm. The laser is passively mode-locked to generate 7 ps FWHM pulses at a repetition rate of 81.25 MHz, this being the 16th sub-harmonic of the main linac frequency. The drive laser optical system [4] defines and delivers macropulses to the cathode at a repetition rate up to 20 Hz. The macropulse length can vary from a single laser pulse up to a 100 μs pulse train. This constitutes a maximum duty cycle of 0.2 % and represents an average electron beam current of 13 μA at an electron bunch charge of 80 pC.

[1] Electron Model of Many Applications non-scaling Fixed-Field Alternating Gradient accelerator

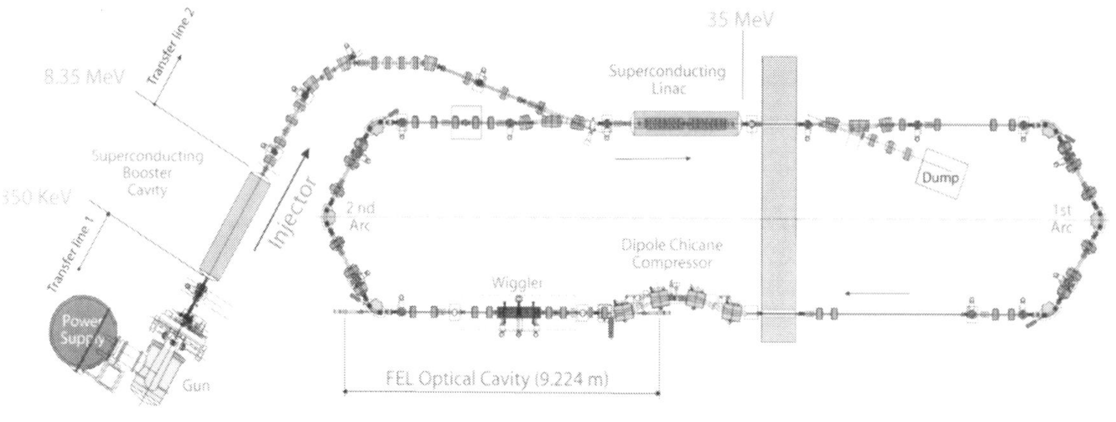

Figure 1. Schematic layout of the ALICE ERL.

The booster linac uses 2 × 9–cell 1.3 GHz TESLA cavities to accelerate the beam to 8.35 MeV, with the main linac using a further 2 cavities to accelerate the beam to 35 MeV. The beam is passed through a compression chicane, and the compressed beam is used to drive the FEL.

3. ALICE Exploitation

The ALICE electron beam is intended primarily to drive an Infra–Red FEL, but in generating the high peak current needed to drive the FEL, the electron beam becomes a source of THz radiation due to coherent enhancement at these wavelengths as it passes through the dipole compression chicane. The intensity of the THz radiation is closely linked to the energy chirp applied by the linac phase, as shown in figure 2. ALICE constitutes the most intense pulsed THz source in Europe, and this radiation is supporting an experimental programme run by the University of Liverpool to investigate the effect of high–power THz radiation on living cells.

Ultra–short X–ray pulses have been generated with the ALICE electron beam through Compton back–scattering[2] where the electron beam was collided head–on with a Ti:sapphire laser pulse (800 nm, 70 fs, 0.5 J) [2], as shown in figure 3. The head–on geometry produces X–

[2] Strictly known as *Thompson scattering* in this low–energy regime

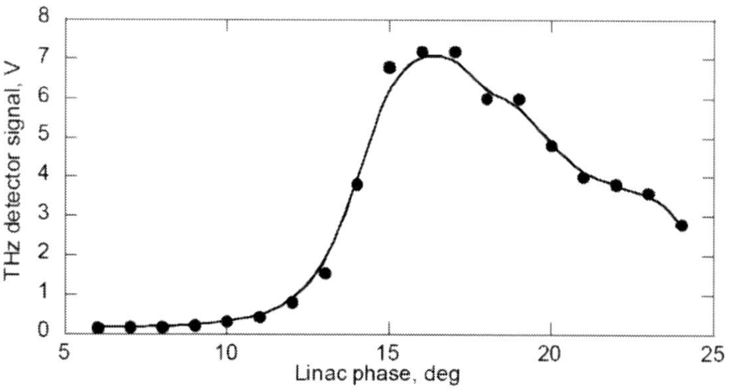

Figure 2. Intensity of the THz emission from the dipole compression chicane as a function of linac RF phase.

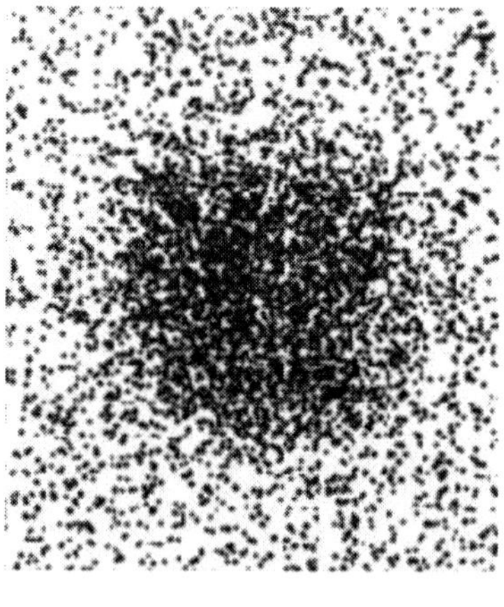

Figure 3. CBS X–ray pulse.

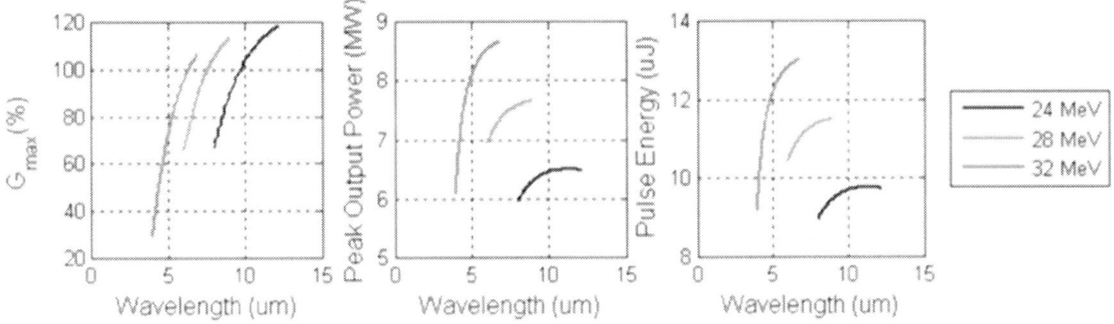

Figure 4. Predicted output characteristics for the ALICE IR–FEL.

rays up to 30 keV with the pulse length limited by the electron bunch length. There remains an intention to re–visit this scheme using a more–challenging side–on collision geometry to generate X–rays pulses up to 15 keV whose pulse length is limited only by the laser pulse length.

The FEL output will be tunable between 4 and 12 μm by varying the electron beam energy between 24 and 35 MeV, and the wiggler gap between 12 and 20 mm. The predicted characteristics are shown in figure 4 where G_{max} is the single–pass gain. Acheiving first lasing of the FEL is the highest priority for the commissioning team[3].

ALICE also serves as the injector for the EMMA accelerator. EMMA is the world's first non–scaling fixed–field alternating gradient electron accelerator, a machine which combines the best features of synchrotrons and cyclotrons [5]. ALICE provides electron bunches to EMMA via an injection line which contains a suite of diagnostics to fully–characterise the incoming beam. The electrons undergo rapid acceleration in EMMA, after which they will be extracted into a diagnostic line with a second suite of diagnostics to characterise the accelerated beam. The NS–FFAG technology has potential applications in the fields of hadron therapy and electricity generation under the accelerator–driven sub–critical reactor scheme. Currently, coasting beams have been circulated, but the beam has not yet been accelerated.

4. Photocathode performance

The GaAs wafer is indium–soldered to the end of a support tube located inside the cathode assembly support stem (shown in figure 5). During activation, the wafer is retracted inside the cathode ball such that it can be exposed to caesium from SAES sources. The oxidant is introduced via a fine leak valve, effectively back–filling the gun chamber. This *in–situ* process also exposes the high voltage surfaces to caesium, a situation which is not ideal and has prompted the development of an external Photocathode Preparation Facility (PPF) [6], as described in section 6. The lack of fine control through the *in–situ* activation limits the maximum achievable *QE*, the best performance so far seen with this system being a *QE* of 3.7% with a 900 hour dark lifetime. This lifetime was achieved despite suffering from a leak into the gun chamber which elevated the base pressure from the mid 10^{-12} mbar regime to the low 10^{-11} regime. The performance of the photoinjector gun was described in detail at the last PESP workshop [1].

5. Injector performance

There have been several recent performance measurements of the ALICE injector line. As a prelude to EMMA beam commissioning, using the EMMA setup, the emittance in the ALICE

[3] Following the PESP workshop in September, first lasing of the ALICE FEL was achieved on 23rd October 2010

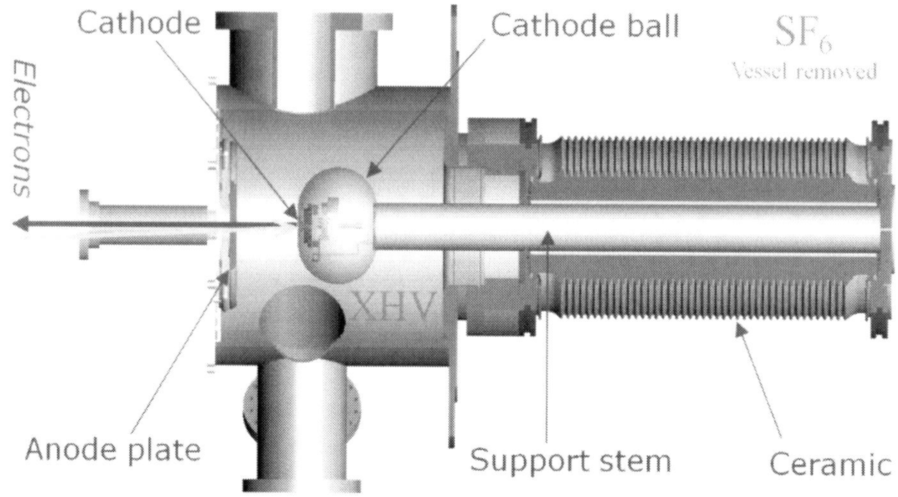

Figure 5. Schematic section showing the photoinjector gun assembly.

injector was measured using a number of techniques [7]. At the time of these measurements, the drive laser beam footprint on the photocathode was sub–optimal, and rather than being circular with a 4 mm diameter, it was elliptical with semi–major and –minor axes of 8 and 4 mm respectively. This served to inflate the emittance significantly.

Working at a beam energy of 6.5 MeV, the emittance was determined using a single *stationary* slit. The beam size after the booster linac was determined from a YAG screen image, then the YAG screen was replaced with a 10 μm slit centred on the beam. Another image was taken from a YAG screen located a distance d further downstream. These measurements showed that:

$$\epsilon_{n,x} = \frac{\sigma_1 \cdot \sigma_2}{d} \cdot \gamma = 11.0\,\text{mm mrad} \tag{1}$$

When the slit was *scanned* to provide a better measurement of emittance, it was found that:

$$\epsilon_{n,x} = \sqrt{\langle x^2\rangle\langle x'^2\rangle - \langle xx'\rangle^2} \cdot \gamma = 9.1\,\text{mm mrad} \tag{2}$$

Quadrupole scans were also carried out. The measured beam size $\langle x\rangle$ at a distance d from each quadrupole was related to normalised emittance as shown in equation (3), where kL is the integrated quadrupole field strength and A, B & C are polynomial curve–fitting constants.

$$\begin{aligned} \langle x^2\rangle &= A(kL)^2 - 2AB(kL) + (C + AB^2) \\ \epsilon_n &= \frac{\sqrt{A \cdot C}}{d^2} \cdot \gamma \\ \epsilon_{n,x} &= 15.1\,\text{mm mrad} \quad \text{and} \quad \epsilon_{n,y} = 4.0\,\text{mm mrad} \end{aligned} \tag{3}$$

These measured values were compared to the values extracted from a GPT [10] simulation. GPT predicted $\epsilon_{n,x} = 9.5$ mm mrad and $\epsilon_{n,y} = 1.9$ mm mrad for the electron beam with a 'perfect' elliptical drive laser beam spot, though this is not in good agreement with the measured values. However, when the GPT simulation was modified such that the initial particle distribution was based on the actual non–uniform intensity distribution extracted from an image of the drive laser spot on the cathode, the values returned were $\epsilon_{n,x} = 17.7$ mm mrad and $\epsilon_{n,y} = 3.8$ mm mrad. These values are in good agreement with the measured values, but fall significantly short of the design specification for ALICE.

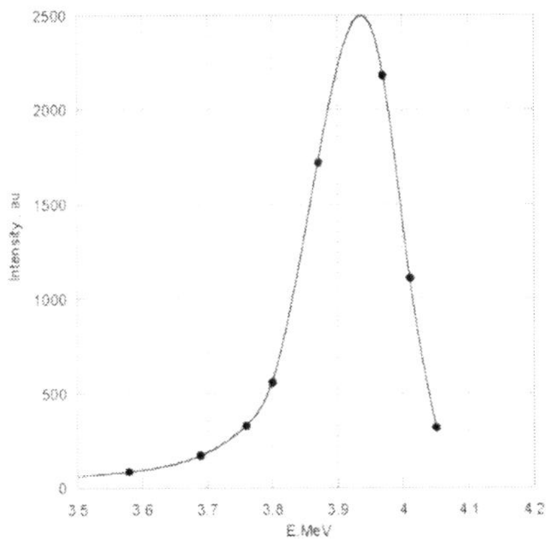

Figure 6. Total energy spread.

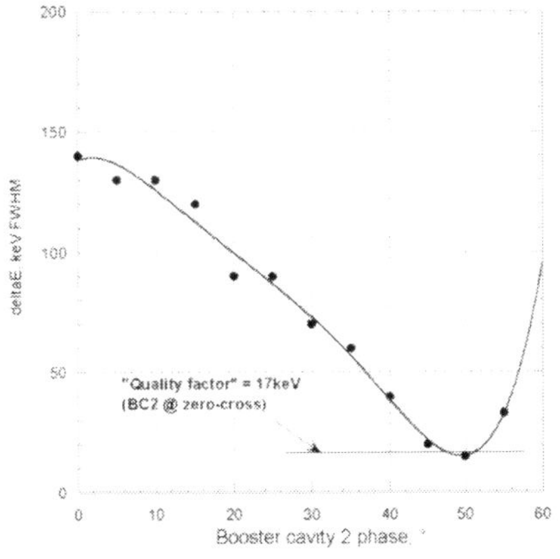

Figure 7. Un–correlated energy spread.

The injector beam energy spread with the FEL setup has been measured using a magnetic spectrometer [8]. With booster cavity 1 operated on–crest to deliver beam at 3.9 MeV, and booster cavity 2 un–powered, the energy spectrum shown in figure 6 was obtained. This corresponds to a total energy spread of 150 keV FWHM. Booster cavity 2 was then powered to accelerate the beam to a final energy of 6.5 MeV, and its phase swept to find the minimum energy spread whilst the gradient was continually adjusted to maintain the beam at 6.5 MeV. This resulted in an un–correlated energy spread of 17 keV at 6.5 MeV, as shown in figure 7.

The '*zero–cross*' method [9] has been applied to bunch length measurement in the injector. Operating ALICE with the FEL setup, it has been found that the bunch length in the injector beamline is 4 mm FWHM, corresponding to a bunch duration of 13.5 ps [8]. The ALICE injector can be configured with shorter bunch lengths, but this is at the expense of the energy spread.

The electro–optic technique has been applied in a separate bunch length measurement *after* the dipole compression chicane. Operating ALICE in the FEL setup, it has been found that the bunch length is less than 1 ps.

6. Photocathode Preparation Facility (PPF)

The ALICE photoinjector will be upgraded during Summer 2011 with the addition of an external three–chamber PPF [6], as shown in figure 8. This upgrade will also require modification of the gun vacuum chamber and the cathode assembly to facilitate the transfer in and out, loading and removal of photocathodes.

A transport vessel detaches from the PPF and is placed inside a nitrogen–purged glovebox for loading with up to 4 photocathodes. Photocathodes made at the Novosibirsk Institute of Semiconductor Physics are un–packed inside the glovebox, etched in HCl, rinsed in iso-propanol, then loaded into the transport vessel. The transport vessel is then sealed, removed from the glovebox and bolted to the PPF loading chamber. The loading chamber is pumped–down, then the gate valve isolating the cleaning chamber is opened. The middle chamber allows cathodes to be heat–cleaned, or previously–operated cathodes to be rejuvenated through exposure to atomic hydrogen (AHC). The preparation chamber is located beyond the cleaning chamber, and isolated by another gate valve. It contains a carousel capable of holding up to 6 photocathodes, a further 2 heat–cleaning stations, and an activation station with access to laser light at a number of different wavelengths. Two magnetic linear transfer arms facilitate movement of cathodes between the 3 PPF chambers, and between the PPF storage carousel and the photoinjector gun.

Attention has been paid to the preparation chamber vacuum environment, and pressures in the mid 10^{-12} mbar regime are normal. Operation of this PPF as a stand–alone facility has become routine since its commissioning in March 2009, and *QEs* of 15 % at 635 nm are common.

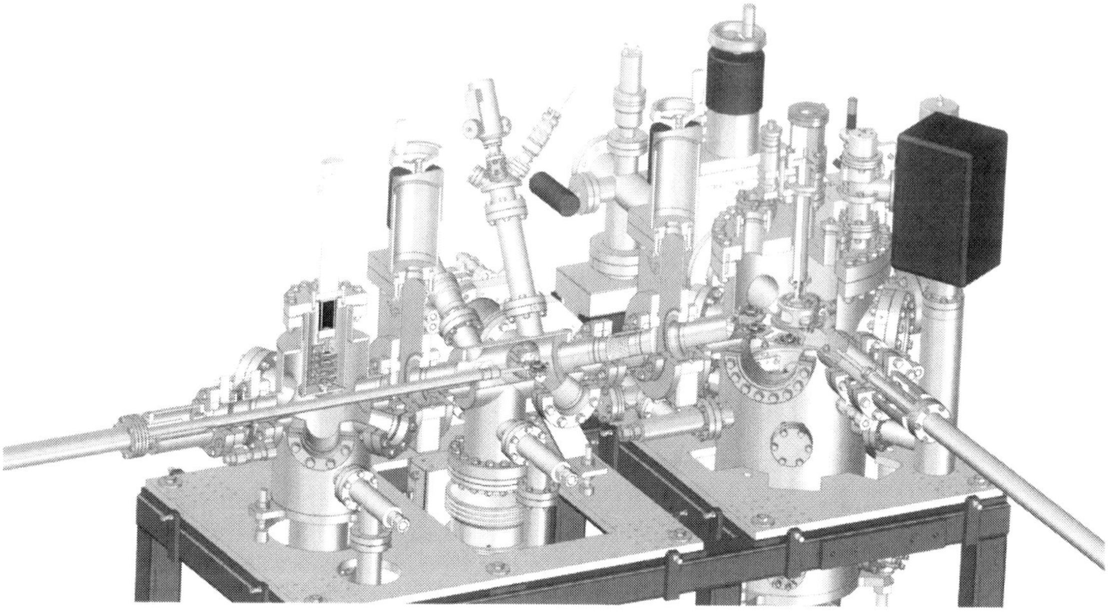

Figure 8. Engineering model of the 3–chamber PPF upgrade for the ALICE photoinjector. **Left–to–right:** Loading chamber; AHC and thermal cleaning chamber; preparation chamber.

7. Summary and future plans

Commissioning of the ALICE ERL has progressed significantly since the last PESP workshop. Operation with energy recovery has become routine, and the machine is supporting a number of scientific programmes, with more to follow. FEL Commissioning is the highest priority, and this will be the focus for the remainder of the final operational period of 2010. Beamlines for extraction of the FEL radiation are currently being installed to support the exploitation programme. Major upgrades planned for the next year include the installation of an up–graded linac cryomodule to support high–current operation at 35 MeV, addition of the PPF, and a gun diagnostic beamline. There is also a plan to replace the current analogue low–level RF system with a digital system for improved cavity phase stability and reduced beam loading.

References

[1] Jones L B, Jamison S P, Saveliev Y M, Middleman K J & Smith S L 2008 *AiP Conf. Proc.* **1149** pp 1084–88
[2] Priebe G *et. al* 2010 *SPIE* (submitted)
[3] Siggins T, Sinclair C K, Bohn C, Bullard D, Douglas D, Grippo A, Gubeli J, Krafft G A & Yunn B 2001 *Nuclear Instruments & Methods A* **475** pp 549–53
[4] Jones L B 2007 *Proc. ERL '07* pp 112–14
[5] Edgecock R 2006 *Nuclear Physics B (Proc. Suppl.)* **155** pp 321–22
[6] Militsyn B L, Burrows I, Cash R J, Fell B D, Jones L B, McKenzie J W, Middleman K J, Scheibler H E & Terekhov A S 2010 *ICFA Beam dynamics newletter* **51** pp 161–70
[7] Garland J, Owen H L, Muratori B D & McKenzie J W 2010 *Proc. IPAC '10* pp 4343–45
[8] Saveliev Y M 2010 *Period 09-10 AP report* (Internal ALICE document)
[9] Wang D X, Krafft G A and Sinlair C K 1998 *Phys. Rev. E* **57**(2) pp 2283–86
[10] van der Geer S B and de Loos M J, General Particle Tracer: a 3D Code for Accelerator and Beam Line Design, EPAC98, www.pulsar.nl/gpt

Operation of the superconducting RF photo gun at ELBE

J Teichert[1,5], A Arnold[1], H Büttig[1], U Lehnert[1], P Michel[1], P Murcek[1], Ch Schneider[1], R Schurig[1], G Staats[1], R Xiang[1], P Kneisel[2], T Kamps[3], J. Rudolph[3], M. Schenk[3], G. Klemz[4] and I Will[4]

[1] Forschungszentrum Dresden-Rossendorf, P O Box 510119, 01314 Dresden, Germany
[2] Thomas Jefferson National Accelerator Facility, 12000 Jefferson Avenue, Newport News, VA 23606, USA
[3] Helmholtz-Zentrum Berlin, Albert-Einstein-Str. 15, 12489 Berlin, Germany
[4] Max-Born-Institut, Max-Born-Str. 2, 12489 Berlin, Germany

E-mail: j.teichert@fzd.de

Abstract. As the first superconducting RF photo-injector (SRF gun) in practical operation, the SRF gun has been successfully connected to the superconducting linac ELBE at Forschungzentrum Dresden-Rossendorf. The injection with this new gun will improve the beam quality for the users of the radiation source. The SRF gun contains a 3½ cell superconducting accelerating cavity with a frequency of 1.3 GHz. The design is for use of normal conducting photocathodes. At present, caesium telluride photocathodes are applied which are illuminated by an ultraviolet laser beam. The kinetic energy of the produced electron beam is 3 MeV which belongs to a peak electric field of 16 MV/m in the cavity. The maximum bunch charge which is obtained and measured in a Faraday cup is about 400 pC (20 μA average current at a repetition rate of 50 kHz). The SRF gun injector is connected to the ELBE accelerator via a dogleg with two 45° deflection magnets. This connection beam line was commissioned in January 2010. A first beam injection into the ELBE accelerator has been carried out with a bunch charge of 120 pC (6 μA at 50 kHz). Detailed measurements showed that beam loss occurred in the dogleg above 60 pC due to the correlated energy spread. In order to find the optimal operation conditions, energy spread was measured in dependence of bunch charge, laser phase and further gun parameters. The Cs_2Te photocathode shows an excellent life time. It is in the gun since May 2010 with about 300 h beam time and about 7 C extracted charge. In the present cavity, the limit for the acceleration gradient is field emission due to some defect on the cavity surface and problems during cleaning. Therefore a modified 3½ niobium cavity has been fabricated, which will increase the RF gradient in the gun and thus improve the beam parameters further.

1. Introduction

Future accelerator based radiation sources need high-brightness electron beams with high average currents and pulse repetition rates in the megahertz rage. Electron guns which can provide beams with these properties are still under development. The most promising candidate is the superconducting photo electron injector (SRF gun). The basics operation principle of a SRF gun is the direct production

[5] To whom any correspondence should be addressed

of short electron pulses by laser irradiation of a photocathode which is placed in a RF resonator. The RF field ensures an instantaneous and strong acceleration of the electron bunches. The superconducting RF resonator has an extremely high quality factor and thus very low thermal losses which allows for a continuous wave (CW) operation. Compared to other CW guns, e.g. thermionic or DC voltage guns, the SRF gun has the highest gradient and acceleration voltages combination. Due to the low RF power losses, nearly all of the RF power is consumed for beam acceleration. And the SRF gun vacuum is ideal with an enormous cryopumping speed to provide best lifetime for sensitive photocathodes. At FZD a SRF gun was developed and commissioned [1] for the ELBE electron accelerator [2]. This gun is installed in the ELBE accelerator hall together with a sophisticated diagnostic beam line [3] and an UV driver laser system [4]. Until end of 2009 the SRF gun had solely been operated simultaneously to the ELBE user operation. The gun was optimized and the diagnostics commissioned. Measurements were carried out with respect to cryogenics, RF properties, photocathodes, and beam parameters and are published elsewhere [5]. During the 2009/2010 winter shutdown, the beam line was installed which connects the SRF gun to the ELBE linac.

2. Photocathodes

2.1. Preparation and exchange

The SRF gun requires photocathodes with quantum efficiency (QE) of 1 % or higher with a stability over weeks. The Cs_2Te photocathodes are chosen as the standard photocathodes. They are prepared in a special cathode lab and then transported to the accelerator hall. The tip of the cathode consists of molybdenum with 99.9 % purity and has 10 mm diameter. The Cs_2Te layer deposited on the top surface has 8 mm diameter. The deposition is done in vacuum between 10^{-9} and 10^{-8} mbar [6]. The QE of fresh photocathodes is between 8 % and 15 %. The cathodes are stored and transported to the SRF gun in a chamber with the vacuum of 2×10^{-10} - 1×10^{-9} mbar. The process of cathode transfer and also the operation in the SRF gun causes a quick decrease of the QE to about 1 %. On this level the QE remains stable and it seems that the photocathodes can serve for months in the SRF gun. A photocathode exchange needs about 30 min taking a fresh cathode from the storage chamber flanged to the gun. The exchange does not need a warm-up of the gun. The transport of the storage chamber from the cathode lab to the SRF gun and its docking at the load-lock needs one maintenance shift and a following one-week bake-out during accelerator operation.

2.2. Quantum efficiency and life time

Up to now five Cs_2Te photocathodes have been employed in the SRF gun. An overview is given in table 1. The first cathodes had poor quality due to insufficient vacuum or they were destroyed by accidentally venting during accelerator shut-downs. The present photocathode #250310Mo has served in the gun since four months with an accumulated beam time of about 300 h and a delivered charge of 6.4 C.

Table 1. Overview of Cs_2Te photocathodes used in the SRF gun.

Cathode no	Serving time	QE in gun	Delivered charge	Beam time
#090508Mo	23/5/08-23/6/08	0.05 %		
#070708Mo	21/7/08-19/9/08	0.1 %		
#310309Mo	08/5/09-24/8/09	1.1 %	0.8 C	~150 h
#040809Mo	24/8/09-23/3/10	0.6 %	0.7 C	~100 h
#250310Mo	25/5/10- in use	0.8 %	6.4 C	296 h

2.3. Thermal emittance

The term "thermal emittance" means the normalized transverse emittance of the electrons when they escape from the cathode surface. In the former thermionic guns, the thermal emittance is determined by the cathode temperature. However, for the semiconductor photocathodes the thermal emittance depends on the energy excess, the electrons possess after absorption of a photon and emission. The normalized initial transverse emittance of a RF gun can be expressed as $\varepsilon_{n,rms}^2 \approx \varepsilon_{therm}^2 + \varepsilon_{sp}^2 + \varepsilon_{rf}^2$ and contains contributions of the thermal emittance ε_{therm}, the space charge part ε_{sp}, and the RF part ε_{rf}. In order to keep the second and third term small, the measurement of the transverse emittance must be carried out at very low bunch charge as well as the RF field should be low and the laser pulse short enough. These conditions are fulfilled in our experiments. A theoretical calculation of the thermal emittance for Cs_2Te cathode has been performed by Floettmann [7] within the three-step model with the result

$$\varepsilon_{therm} = \frac{x_{rms}}{\sqrt{3}}\sqrt{\frac{2E_{kin}}{m_0 c^2}} \quad . \tag{1}$$

In the formula x_{rms} is the rms laser spot size on the photocathode and E_{kin} is the electron kinetic energy given by $E_{kin} = E_{photon} - E_G - E_A$ with the photon energy E_{photon}, the band gap E_G, and the electron affinity E_A. By variation of the laser spot size, E_{kin} can be determined experimentally:

$$E_{kin} = \frac{3}{2} m_0 c^2 \left(\frac{d\varepsilon_{therm}}{dx_{rms}}\right)^2 \quad . \tag{2}$$

The wave length of the UV laser is 262nm, corresponding to the photon energy of 4.72 eV. The Cs_2Te semiconductor at liquid nitrogen temperature has the same band structure as the one at room temperature. According to Powell's measurement [8], the electrons will be excited to the final state in the conduction band with 4.05 eV above the valence band maximum. The band gap is 3.3eV, and the electron affinity is 0.2 eV for a fresh Cs_2Te photocathode film without external electric field. In such case, the electron kinetic energy can be calculated as E_{kin}= (4.05-3.3-0.2) eV = 0.55 eV. But for a polluted emission surface the electron affinity will be changed. Normally the absorption of rest gas will increase the electron affinity, which results in the degradation of QE.

The transverse emittance was measured versus rms laser spot size for different photocathodes with the results shown in figures 1and 2. The laser spot size was adjusted by changing the aperture diameter. The transverse emittance was determined by means of the solenoid scan method using the emittance compensation solenoid and the second YAG screen in the diagnostic beam line.

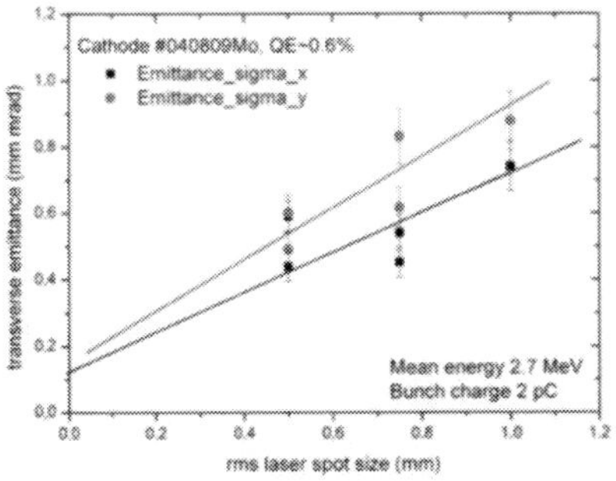

Figure 1. Emittance measurement versus laser spot size for cathode #040809Mo with QE = 0.6%.

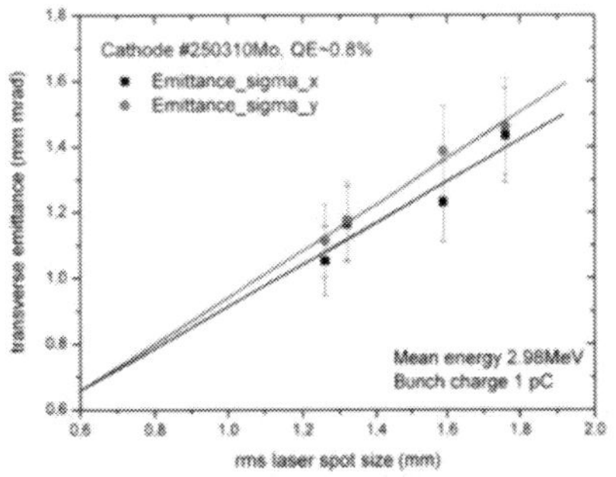

Figure 2. Emittance measurement versus laser spot size for cathode # 250310Mo with QE = 0.8%.

For cathode #040809Mo in figure 1, the slope of $\varepsilon_{rms,x}$ is 0.6 mrad, that of $\varepsilon_{rms,y}$ is 0.78 mrad. From equation 2, the average kinetic energy of the emitted photoelectrons E_{kin} was calculated to (0.38±0.1) eV, which means that the electron affinity is E_A= (0.37±0.1) eV. For cathode #250310Mo in figure 2, the slope of $\varepsilon_{rms,x}$ is 0.64 mrad, that of $\varepsilon_{rms,y}$ is 0.71 mrad. The average kinetic energy E_{kin} was (0.35±0.04) eV, the electron affinity is E_A= (0.40±0.04) eV, about 0.2eV higher than that of the fresh cathode.

2.4. Multipacting

Multipacting is a well-known phenomenon frequently observed in RF systems and may cause serious problems. An analysis of the cavity, choke filter cell, and the other RF components during the design stage of the SRF gun showed a risk for multipacting in the opening where the cathode reach into the half-cell of the cavity. The RF field across the circular gap can accelerate electrons from one surface toward the other. Upon impact further electrons can be released. If the secondary emission coefficient is high enough and the field is resonant this process can continue and lead to multipacting. For that reason the photocathode can be biased with a high voltage to destroy the resonant condition. In fact, we observe this phenomenon during ramping up the RF field in the cavity and need up to 7 kV at the cathode to suppress it.

Figure 3. Modified photocathode stem with grooves for multipacting suppression.

Since the voltage needed varies more or less randomly, we looked for a method to prevent multipacting by another way. At BNL it was tried to suppress multipacting in the quarter wave choke filter by grooves [9] and a detailed study of multipacting suppression by fine grooving of a coaxial input coupler is presented in Ref [10]. In a similar way grooves has been machined in a cathode stem as it is shown in figure 3. In the critical region the cathode has wavelike grooves with 1 mm radius.

The success of the modified cathode geometry is presented in the time diagrams in figures 4. The curve in figure 4a shows the RF curve (gradient versus time) of a standard photocathode. The spikes indicate the appearance of cathode current due to multipacting. After some time the RF was turned off and then the ramp was started again with a higher DC voltage at the cathode. The curve in figure 4b shows the measurement for the grooved cathode. Here the multipacting is completely suppressed. Spikes occur neither for the slow RF ramp nor the following fast ramp.

3. SRF Gun Cavity and Beam Parameters

The quality degradation of the superconducting cavity due to pollution from the normal conducting photocathode in a SRF gun is still a question in discussion. For that reason the cavity performance has been measured regularly. The standard method is to determine the He consumption as function of the cavity gradient. From these data the unloaded quality factor can be calculated as it is shown in figure 5. The cavity can be operated until a peak field of 16 MV/m which corresponds to 6 MV/m acceleration gradient. After more than 650 h beam time, there is still no change in the cavity performance. After each shut-down with a cryomodule warm-up, the cavity was slightly worse. But a short high-power RF processing (HPP) reproduces the old values.

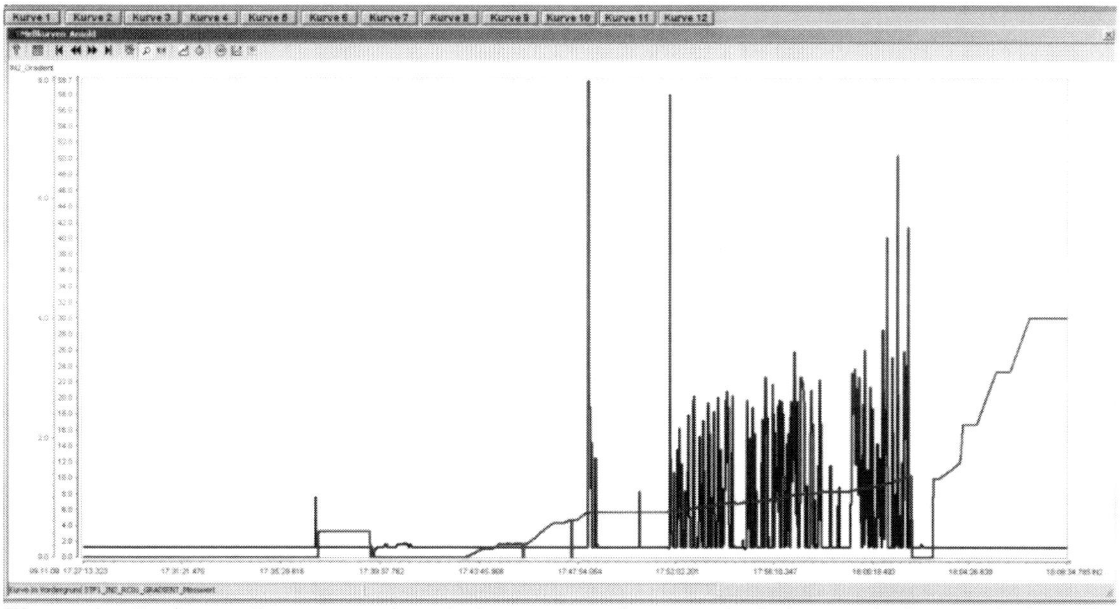

Figure 4a. Cavity gradient versus time during ramp-up for a standard photo cathode stem design. The spikes in the curve are multipacting events.

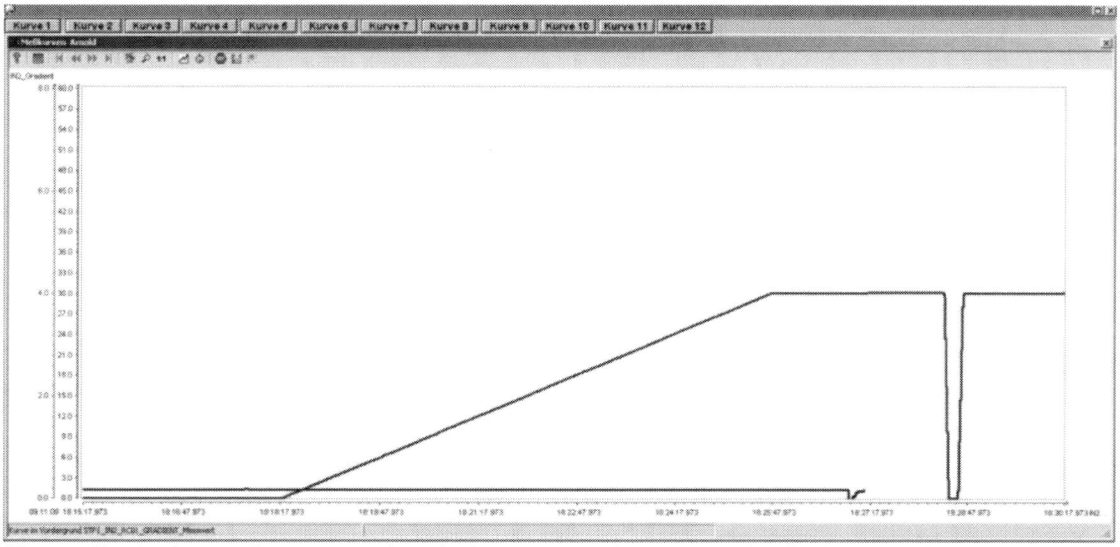

Figure 4b. Cavity gradient versus time during ramp-up for a grooved cathode stem. The curve does not show any multipacting events.

The SRF gun was operated with an acceleration gradient of 6 MV/m always in CW mode. Additionally a DC voltage of -5 kV was applied to the photocathode which slightly improves the beam parameters. The 263 nm laser beam was delivered by a Nd:YLF laser system with a 26 MHz oscillator, a regenerative amplifier, and frequency quadrupling. The average output power maximum was about 300 mW. The pulse repetition rate was chosen to 125 kHz (variable between 1 kHz and 250 kHz). The laser pulse is Gaussian in time with about 15 ps FWHM and had a nearly Gaussian lateral profile with 3.5 mm FWHM. The final kinetic energy and energy spread versus laser phase measured in the diagnostic beam line are presented in figures 6 and 7, respectively. For the SRF gun it is typical, that the electron energy has a broad maximum at low laser injection phases. The energy spread, mainly correlated energy spread, is small for low laser phases and increases with higher laser phases and higher bunch charges. Since a low energy spread is important for the transmission in the dogleg, a low laser phase of 10° was used in the experiments.

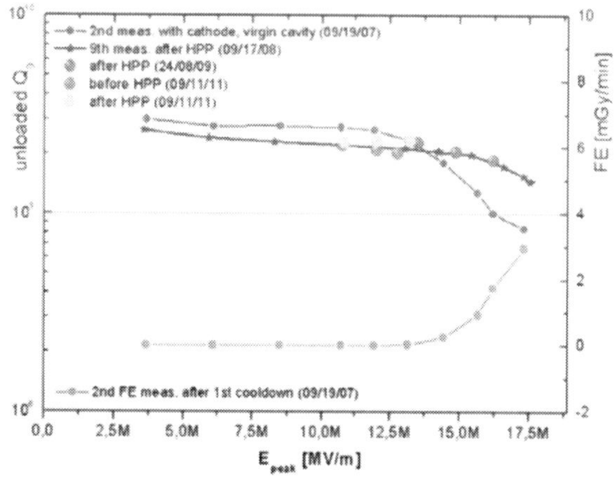

Figure 5. Performance of the SRF gun cavity. The diagram shows the unloaded quality factor versus RF peak field for several measurements and the corresponding field emission (FE).

The SRF gun was operated with an acceleration gradient of 6 MV/m always in CW mode. Additionally a DC voltage of -5 kV was applied to the photocathode which slightly improves the beam parameters. The 263 nm laser beam was delivered by a Nd:YLF laser system with a 26 MHz oscillator, a regenerative amplifier, and frequency quadrupling. The average output power maximum was about 300 mW. The pulse repetition rate was chosen to 125 kHz (variable between 1 kHz and 250 kHz). The laser pulse is Gaussian in time with about 15 ps FWHM and had a nearly Gaussian lateral profile with 3.5 mm FWHM. The final kinetic energy and energy spread versus laser phase measured in the diagnostic beam line are presented in figures 6 and 7, respectively. For the SRF gun it is typical, that the electron energy has a broad maximum at low laser injection phases. The energy spread, mainly correlated energy spread, is small for low laser phases and increases with higher laser phases and higher bunch charges. Since a low energy spread is important for the transmission in the dogleg, a low laser phase of 10° was used in the experiments.

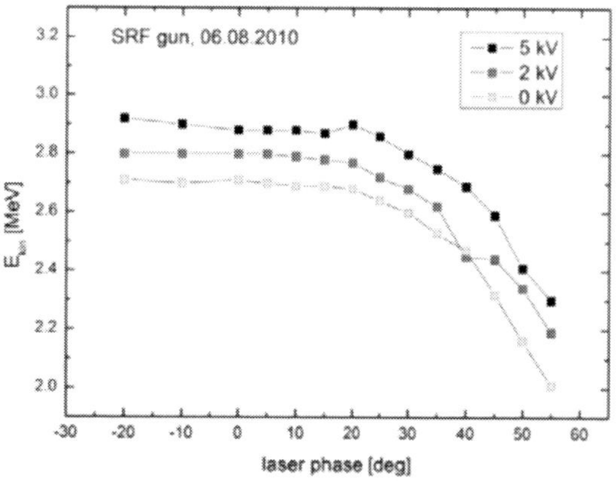

Figure 6. Measurement of the kinetic energy as function of laser phase for an acceleration gradient of 6 MV/m and different DC bias at the cathode.

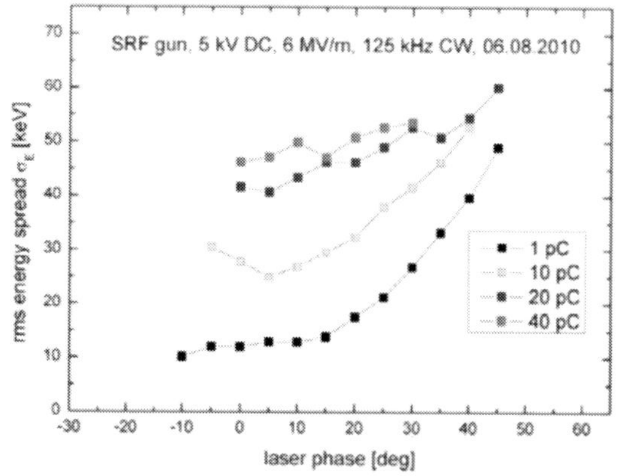

Figure 7. Energy spread at the SRF gun as function of laser phase for different bunch charges at an acceleration gradient of 6 MV/m and -5 kV cathode bias.

4. Beam for ELBE Linac

The beam line layout of SRF gun and ELBE linac is presented in figure 8. The beam line (dogleg), recently installed, connects the SRF gun injector with the ELBE linac. Usually ELBE is operated with the thermionic injector. Downstream the SRF gun a solenoid for emittance compensation is placed followed by a screen station, a movable Faraday cup for current measurements, and a quadrupole triplet. The new beam line starts about 1.5 m from the SRF gun exit with a 45° dipole magnet, which deflects the beam towards the ELBE linac, followed by a quadrupole triplet and ends with a second 45° magnet in front of the first acceleration module of ELBE.

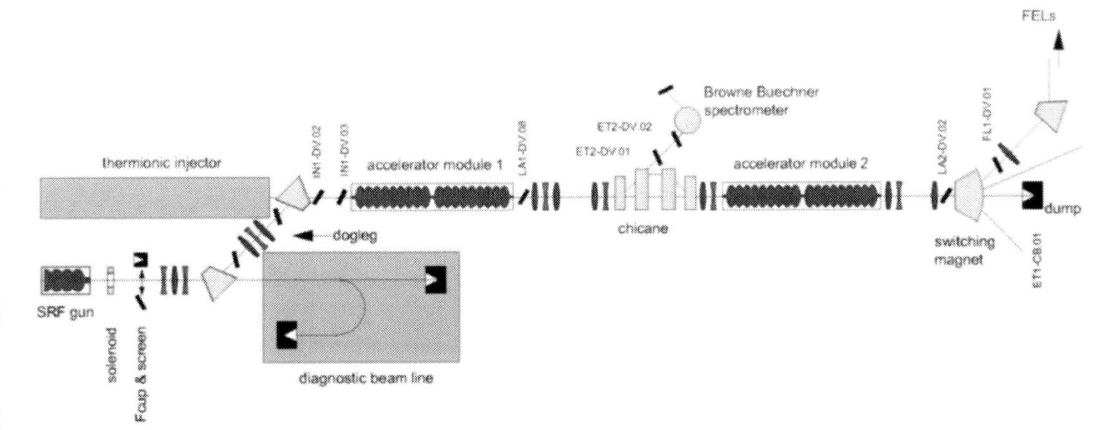

Figure 8. beam line layout of the SRF gun, dogleg section, and ELBE main accelerator.

The injection into the ELBE linac was carried out step by step, increasing the bunch charge. At low bunch charges the quadrupole triplet in the dogleg was adjusted towards an achromatic beam transport and maximum acceptance. After that, the on-crest phases of the first acceleration module were determined searching for maximum energy using the first chicane dipole. The first cavity was used to compensate the correlated energy spread of the gun, as it is shown in figure 9. The on-crest phase (maximum energy) in this measurement was found to be at 28°. The lowest energy spread turned out to be about -10° away. Finally the on-crest phases of the cavities of the second module were determined by means of energy measurement after the switching magnet.

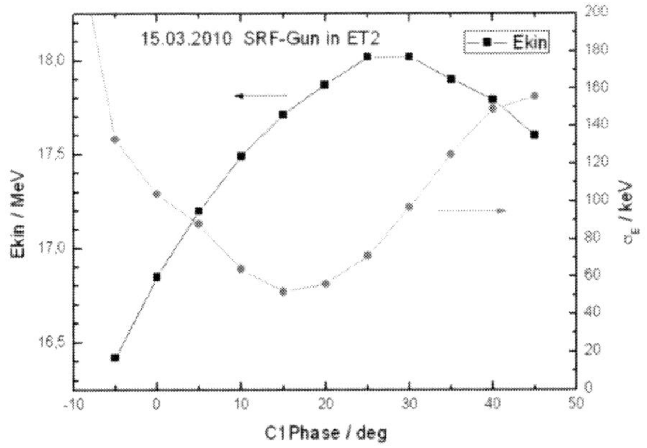

Figure 9. Measurement of energy and rms energy spread after the acceleration module 1 using the first dipole magnet of the chicane and YAG screen ET2-DV.01. The phase of the first cavity C1 was varied and the second cavity phase was held on crest.

Unfortunately, the large correlated energy spread of the SRF gun can be compensated only in the first cavity. Therefore the beam transport through the dogleg section is difficult. In a first test, a beam with

120 pC (6 μA at 50 kHz) was delivered to ELBE. The result of a second injection test is presented in figure 10. The transmission was obtained by comparing the beam currents measured in the Faraday cup near the SRF gun and in the beam dump downstream the switching magnet. Another indicator was the beam-loss monitoring system. The results of the current measurements are presented in figure 10. Up to 7.5 μA beam current (60 pC) the transmission was measured to be equal to one within measurement accuracy and no beam loss was monitored. For higher currents we found a strongly increasing beam loss in the dogleg section, which needs further optimization of the beam transport and gun parameters.

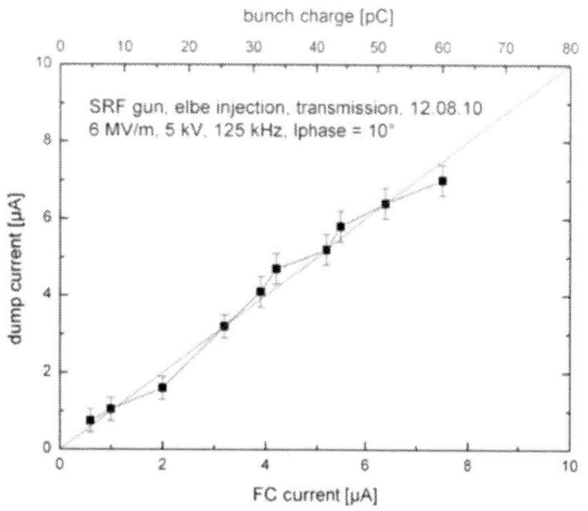

Figure 10. Transmission measurement of the SRF gun injection experiment. The diagram compares the current measured in the Faraday cup downstream the gun with the current measured in the beam dump ET1-CB.01after the switching magnet.

5. Summary

The SRF gun is now connected to the ELBE linear accelerator. The operation of the SRF gun as injector for ELBE has been demonstrated. The Cs_2Te photocathodes used in the gun showed sufficient quantum efficiency and long lifetime. Up to now, we did not found any cavity degradation, which may occur due to the operation of a normal conducting photocathode in the superconducting cavity. The handicap of the gun is the low acceleration gradient of 6 MV/m obtained with the present cavity. It has a strong influence on the beam parameters. Nevertheless, the gun is able to provide beam for ELBE and will be tested for user operation. In order to overcome the low-gradient problem, a new large grain 3½ cell niobium cavity with optimized shape has been designed, fabricated and will be tested soon.

Acknowledgements

We acknowledge the support of the European Community-Research Infrastructure Activity under the FP7 programme (EuCARD, contract number 227579), as well as the support of the German Federal Ministry of Education and Research grant 05 ES4BR1/8.

References

[1] Arnold A, Büttig H, Janssen D, Lehnert U, Michel P, Möller K, Murcek P, Schneider Ch, Schurig R, Staufenbiel F, Teichert J, Xiang R, Kamps T, Lipka D, Marhauser F, Lehmann W D, Stephan J, Volkov V, Will I and Klemz G 2007 *Nucl. Instr.and Meth. A* **577** 440

[2] Michel P, Büttig H, Gabriel F, Helm M, Lehnert U, Schneider Ch, Schurig R, Seidel W, Stehr D, Teichert J, Winnerl S and Wünsch R 2006 *Proc. International Free Electron Laser Conference FEL'06 (Berlin, Germany, 27 August - 01 September 2006)* pp 488-491

[3] Kamps T, Arnold A, Boehlick D, Dirsat M, Klemz G, Lipka D, Quast T, Rudolph J, Schenk M, Staufenbiel F, Teichert J and Will I 2008 *Review of Scientific Instruments* **79**, 093301

[4] Will I, Klemz G, Staufenbiel F and Teichert J 2006 *Proc. International Free Electron Laser Conference FEL'06 (Berlin, Germany, 27 August - 01 September 2006)* pp 564-566

[5] Teichert J, Arnold A, Büttig H, Janssen D, Justus M, Lehnert U, Michel P, Murcek P, Schamlott A, Schneider Ch, Schurig R, Staufenbiel F, Xiang R, Kamps T, Rudolph J, Schenk M, Klemz G and Will I 2008 *Proc. International Free Electron Laser Conference FEL'08 (Gyeongju, Korea, 24-29 August 2008)* pp. 467-472

[6] Xiang R, Arnold A, Buettig H, Janssen D, Justus M, Lehnert U, Michel P, Murcek P, Schamlott A, Schneider Ch, Schurig R, Staufenbiel F and Teichert J 2010 *Phys. Rev. ST Accel. Beams* **13** 043501

[7] K. Floettmann 1997 *TESLA-FEL report 1997-01, DESY*

[8] Powell R A, Spicer W E, Fisher G B and Gregory P 1973 *Physical Review B* **8** 3978

[9] Burrill A, Ben-Zvi I, Cole M, Rathke J, Kneisel P, Manus R and Rimmer R 2007 *Proceedings of PAC07 (Albuquerque, USA, 25-29 June 2007)* pp. 2544-2546

[10] Abe T, Kageyama T, Sakai H, Takeuchi Y and Yoshino K 2010 *Phys Rev. ST Accel Beams* **13** 102001

Status and perspectives of superconducting radio-frequency gun development for BERLinPro

T Kamps[1], W Anders[1], A Frahm[1], J Knobloch[1], O Kugeler[1], J Rudolph[1], A Matveenko[1], A Neumann[1], T Quast[1], M Schenk[1], M Schuster[1], P Kneisel[2], J Sekutowicz[3], R Nietubyc[4], I Will[5]

[1] Helmholtz-Zentrum Berlin, Albert-Einstein-Strasse 15, 12489 Berlin, Germany
[2] Thomas Jefferson National Accelerator Facility, 12000 Jefferson Avenue, Newport News, VA 23606, USA
[3] Deutsches Elektronen-Synchrotron, Notkestrasse 85, 22607 Hamburg, Germany
[4] Andrzej Soltan Institute for Nuclear Studies, 05400 Otwock-Świerk, Poland
[5] Max-Born-Institut, Max-Born-Str. 2, 12489 Berlin, Germany

E-mail: kamps@helmholtz-berlin.de

Abstract. As part of the BERLinPro study, HZB is developing an SRF photoelectron injector. The R&D will be carried out in three stages, the first of which is currently being installed at HZB's HoBiCaT facility. It consists of an SRF cavity with SC solenoid, electron beam diagnosics and drivelaser systems.

1. Introduction

Helmholtz-Zentrum Berlin is the result of a merger between two research labs, HMI and BESSY, in the Berlin area. HZB build and operates several accelerator facilities for research with photons, neutrons and protons. Due to its expertise in the development of large scale accelerator facilities HZB is ideally suited to realize new accelerator concepts. Therefore HZB will build a demonstrator for the Energy Recover Linac (ERL) paradigm [1]. The ERL was proposed in 1965 by M. Tigner [2] und is possible today with the advent of superconducting RF technology for particle accelerators.
In order to demonstrate generation and energy recovery of a 100 mA average current and 1 mm mrad normalized transverse emittance beam HZB is currently working on BERLinPro, a fully integrated ERL test facility including all major systems found in ERLs like electron gun, booster section, merger beamline, main linac, return loop and high power beam dump.
The ultimate performance of the ERL depends on the ability of the electron source to deliver a high brightness, high-average current electron beam to the main accelerator complex. For maximum flexibility, the source must also be able to generate pulses of higher charge at lower repetition rates to meet specific experimental needs. For flexibility and performance, photoelectron injectors (or short photo-injectors) are the primary candidates. Superconducting radio-frequency (SRF) injectors have the highest potential as electron sources for ERLs, as they are able to operate at 100% duty factor and can generate significantly higher fields than (CW-operated) normal conducting (NCRF) systems. Hence SRF systems can fulfill the ERL injector requirements and provide for maximum flexibility (compared with DC sources). Most importantly, they offer the most potential for continued future improvements, so vital to the upgrade of ERL facilities.

2. Roadmap for photo-injector development

The photo-injector has three major subsystems: The cathode, the laser system to release the electrons and the initial acceleration. For the cathode, important considerations are the thermal emittance, the response time on the sub-ps scale, the photon wavelength for reasonable quantum efficiency and its longevity. The laser system must be able to provide the average and peak powers to produce high currents (of order 100 mA) at bunch charges on the order of 100 pC. At the same time, flexible pulse shaping must be included to minimize the impact of space charge on the emittance. These aspects favor longer-wavelength laser systems (e.g., 528 nm) and will impact the choice of cathode materials. For acceleration, it is vital that both a high field (for rapid acceleration to limit the impact of space charge) as well as a high voltage (to minimize the beam degradation in the subsequent drift space) can be applied.

We propose a multi-cell SRF gun operating at the ERL main linac frequency (1.3 GHz) with an embedded alkali antimonide photocathode. The number of cells is limited by the power capability of the input coupler, but should deliver at least 1 MeV/c of momentum to the photoelectrons combined with a launch field of at least 10 MV/m to overcome space-charge limited emission.
The main challenges for the photo-injector system are:

- Verification of the emittance compensation scheme, optimization of solenoid location (inside or outside the cryogenic vessel of the SRF cavity) with respect to emittance performance and SRF cavity operation.
- The cathode/cavity interface, operation of a room-temperature cathode inside an SRF cavity with high accelerating field (with vacuum shield between cavity and cathode insert and RF choke), achievement of high cathode lifetime and development of a procedure for cathode changing during beam operation.
- Design of minimum beam-disruptive high power RF feeds (for power levels up to 200 kW) and HOM output coupler.

All these issues will be addressed with BERLinPro at HZB. The approach ist staged, tackling one item after the other. The goals and parameters for each stage are summarized in Table 1.

Table 1. Tentative parameters of the roadmap guns.

Parameter	**Stage A** **HoBiCaT gun**	**Stage B** **SourceLab**	**Stage C** **BERLinPro**
Goal	Beam demonstrator	Brightness R&D gun	High average current Production gun
Repetition rate	30 kHz	52 MHz	1.3 GHz
Cathode material	Pb	CsK_2Sb	CsK_2Sb
Drive-laser wavelength	258 nm	526 nm	526 nm
Bunch charge	17 pC	77...200 pC	77 pC
Average current	0.5 μA	5...11 mA	100 mA
Exit energy	0.5...2.5 MeV	0.5...2.5 MEV	1.5 MeV

The stage A of the R&D programme aims at setup and commissioning of an all superconducting high brightness gun. Here the cathode is Pb film which is plasma-arc deposited on the backplane of the cavity. By utilizing the backplane as the photocathode additional complications associated with introducing a normal-conducting photocathode are avoided.
All issues connected with the introduction of a normal-conducting cathode film or stock into the SRF cavity are mitigated to stage B of the programme. For BERLinPro, the baseline cathode material is currently CsK_2Sb, due to the high QE achievable at visible wavelength [3]. Intense activity with regards to cathode preparation and characterization will be unfolded before suitable high QE cathode can be introduced to the gun in Stage B. For the last stage C the goal is to accelerate 100 mA of current to 1.5 MeV, so the development of suitable high average power RF couplers is the main focus of R&D activities in this stage.

3. SRF cavity development
The cavity was designed at HZB based on a microwave model by J. Sekutowicz. The cavity itself was fabricated and tested at Jefferson by P. Kneisel. As the cavity was designed without a tuning scheme special emphasis was laid on the mechanical design to lower the detuning by deformation of the cavity, such improve the expected field stability and thus allow meaningful studies of the beam parameters. The backplane stability is supported by a stiffening spider, long flat bars attached to the backplane.

Figure 1. Cavity parts (left three photos) and fully assembled cavity (right).

4. Pb cathode deposition
The Pb cathode is deposited on the large-grain Nb half-cell back-wall by means of plasma arc deposition. At the setup at the Andrzej Soltan Institute the cavity vessel is installed in a 30 degree angle to the plasma source in order to avoid the formation of droplets while still allowing a high Pb deposition efficiency [4]. After up to 16 cycles of about 5 minute deposition time with 30 minutes pause in between a Pb cathode with a thickness of about 0.5 μm and 3 mm diameter has been formed. During the pauses between deposition cycles the cavity is given time to cool down. The spot is located within a single-crystal lattice domain to have isotropic thermal contact between Pb film and backplane. After deposition and visual inspection of the film, the cavity volume is filled with Ar gas. Then the cavity is sent back to Jefferson Lab for the final BCP and high pressure rinsing treatment, following final vertical tests and assembly of the RF input and pickup couplers.

5. Summary of vertical tests at JLAB

The cavity was tested several times during construction, cleaning and before and after deposition of the Pb film at the vertical test area (VTA) and Jefferson Lab [5]. The cavity was first tested after assembly and welding of the cells, tuning and BCP-high pressure water rinsing treatment. The resulting quality factor Q versus peak electric E_{peak} data is shown in figure 1.

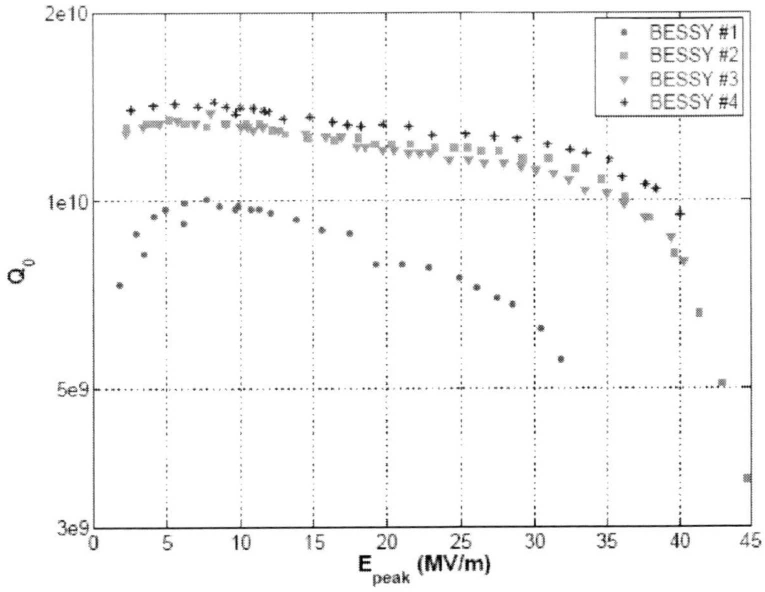

Figure 2. Selected cold RF test results. See text for explanation.

During test #1 the field flatness between half and full cell was only 66%. Further tuning and another BCP treatment resulted in an improvement of the field flatness to 94% and quality factor Q_o exceeding 10^{10} levels for peak electric fields $E_{peak} \leq 35$ MV/m (see test #2). Test #3 was done after mounting of the LHe vessel. There was no significant change in performance noticed. The performance drops in Q_o for field levels above 35 MV/m due to slight field emission. After this test the cavity was sent to A. Soltan Institute for deposition of the Pb film. After deposition and visual inspection the cavity was sent back to Jefferson Lab for final treatment steps. Visual inspection of the Pb spot there showed a yellowish color of the Pb film hinting at contamination with Pb_yO_x. During further treatment steps the film was accidentally lost. Again, the cavity was sent to A. Soltan Insitute for coating. After another round of BCP and high pressure rinsing test #4 indicated partial recovery of the performance.

6. Additional systems

The cavity will be installed together with a superconducting solenoid in the cryovessel of the HoBiCaT facility. HoBiCaT is a test facility at HZB including a complete range of infrastructure necessary for performance tests of superconducting RF systems and cavities [6]. The electron beam diagnostics beamline has all elements required to measure the projected beam emittance, monitor charge transmission and measure the energy and energy spread of the bunches from the SRF gun. The beamline elements are derived from the diagnostics setup for the FZD/ELBE SRF photo-injector [7].

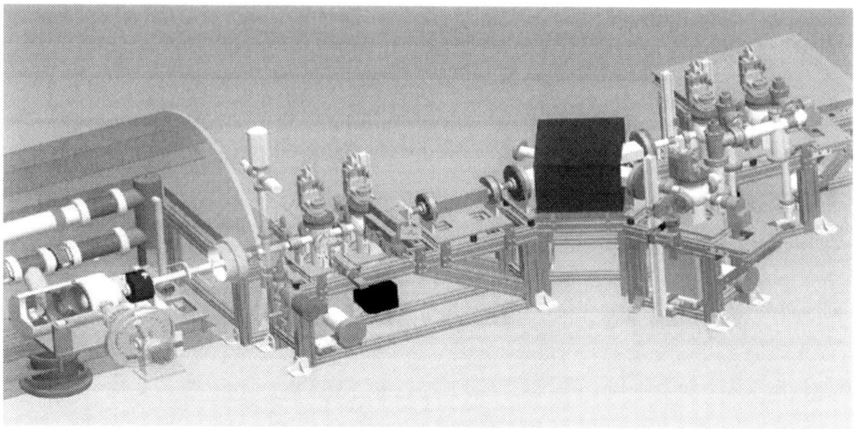

Figure 3. Setup of the gun cavity in the HoBiCaT test facility with electron beam diagnostics beamline.

The photocathode drive-laser is developed by Max Born Institut (MBI) and consists of a diode-pumped Yb:YAG oscillator with a diode-pumped regenerative amplifier. We want to use the 4th harmonic of YAG to illuminate the Pb photocathode with 258 nm laser photons. The quantum efficiency of Pb at this wavelength is close to $5*10^{-4}$ [8]. The laser pulse length will be 2...3 ps FWHM, with pulse energy of 0.15 μJ. The repetition rate is 30 kHz.

7. Outlook

First beam from the all superconducting gun in HoBiCaT is expected for early Spring 2011. In parallel cathode preparation and characterization facilities are setup to be ready by 2013 to start operation of a SRF gun cavity with NC cathode insert and high QE photocathode.

Acknowledgements

The R&D programme for the photo-injector for BERLinPro is supported by Helmholtz-Zentrum Berlin (HZB), the state of Berlin and the Helmholtz Gemeinschaft (HGF).
The work on the Pb cathode is supported by EuCARD Grant Agreement No. 227579.

References

[1] Abo-Bakr A et al. 2010 *Proc. LINAC 2010*, Tsukuba, Japan
[2] Tigner M 1965 *Nuovo Cimento* **37** 1228
[3] Dowell D H, Bethel S Z, Fridell K D 1995 *NIM* A **356**
[4] Nietubyc R et al. 2010 *Proc. LINAC 2010*, Tsukuba, Japan
[5] Kneisel P, to be published
[6] Kugeler O, Neumann A, Anders W et al. 2010 *RSI* **81** 7 074701
[7] Kamps T et al. 2008 *RSI* **79** 9 093301
[8] Smedley J, Rao T, Sekutowicz J 2008 PRST-AB 11 013502

Journal of Physics: Conference Series **298** (2011) 012010 doi:10.1088/1742-6596/298/1/012010

Superlattice Photocathode for High Brightness and Long NEA-surface lifetime

T Nishitani[1], M Tabuchi[2], K Motoki[3], T Takashima[3], A Era[3], Y Takeda[4]
[1] Radiation laboratory, RIKEN, 2-1 Hirosawa, Wako, Saitama, 351-0198, Japan
[2] Venture Business Laboratory, Nagoya Univ., Furo-cho, Chikusa-ku, Nagoya, Aichi, 464-8601, Japan
[3] Department of Materials Science and Engineering, Nagoya Univ., Furo-cho, Chikusa-ku, Nagoya, Aichi, 464-8601, Japan
[4] Department of Crystalline Materials Science, Nagoya Univ., Furo-cho, Chikusa-ku, Nagoya, Aichi, 464-8601, Japan

E-mail: tnishitani@riken.jp

Abstract. We have suggested using a superlattice semiconductor instead of a bulk semiconductor as an NEA-photocathode for small emittance and high quantum yield because the mini-band energy width and high joint density of states at the bottom of the conduction band can bring small photoelectron's momentum spread and high quantum yield at excitation energy of the band gap. We developed the test sample of the AlGaAs-GaAs superlattice semiconductor designed by using Klonig-Penny-Bastard model. The obtained quantum yield spectrum from the AlGaAs-GaAs superlattice semiconductor after NEA-surface activation was a step function of excitation photon energy most likely caused by the quantum confinement effect.

1. Introduction

Photocathodes using semiconductors made of GaAs and related materials with negative electron affinity surfaces (NEA-GaAs) have high quantum yield and can emit spin-polarized electrons excited by a circular polarized laser. NEA-GaAs photocathodes using semiconductors with a strained-layer structure or strained-layer superlattice structure [1] [2] [3] have played a very important role as highly spin-polarized electron sources in high-energy physics [4]. An NEA-GaAs with a bulk structure has also been used as a photocathode which can extract large photocurrent (9mA) for a free electron laser accelerator [5].

The key phenomenon of photocathodes using semiconductors made of GaAs and related materials with negative electron affinity surfaces (NEA-GaAs) is that the NEA surface enables excited electrons in the bottom of the conduction band to escape to a vacuum. However, the surface of conventional NEA-GaAs photocathodes is damaged by back bombardment of ionized residual gas by photoelectrons [6]. The decrease in quantum yield due to the damage to the surface results in a short lifetime during high-photocurrent operation. The conventional NEA-GaAs photocathode with high-brightness beam performance uses a GaAs semiconductor with a bulk structure. However this photocathode has a trade-off in beam performance between the small energy spread of electrons and high photocurrent, because it requires higher-energy photons to achieve high-photocurrent operation at the expense of a larger energy spread of electrons.

We considered that the optimum electron affinity of a semiconductor is essential for improvement of the decrease in quantum yield due to damage to the surface. To confirm this hypothesis, we developed AlGaAs photocathodes with various Al fractions. Those AlGaAs samples achieved a 10-times-longer quantum yield life than the GaAs sample, and had longer quantum yield life by larger Al fraction [7].

2. Superlattice Photocathode

Semiconductors for a highly spin-polarized electron source require that the splitting energy between the heavy- and light-hole energy levels (hh-lh splitting energy) in the valence band is as high as possible. Superlattice photocathodes with a large hh-lh splitting energy in the valence band are known to be a spin-polarized electron source with polarization above 50% [1]. We also considered that semiconductors with a superlattice structure convey the advantages not only for highly spin-polarized electron sources but also for high-brightness electron sources.

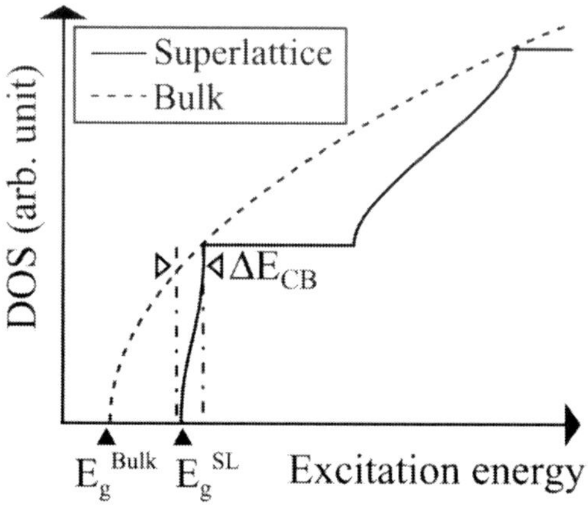

Figure 1. Dependence of DOS on excitation energy for bulk and superlattice structures.

We consider that the superlattice semiconductor has intrinsic advantages for an NEA-GaAs photocathode with higher quantum yield, smaller energy spread of photoelectrons and longer NEA lifetime than conventional NEA-GaAs photocathodes owing the following two factors. (1) The electron energy density of state (DOS) for a superlattice is a step function of excitation photon energy due to the quantum confinement effect as shown in Figure 1. In a NEA-GaAs photocathode, the quantum yield is proportional to the ratio of DOS integrated from the band-gap energy to the excitation photon energy [8]. A GaAs semiconductor with a bulk structure, which has a monotonically increasing DOS function, requires higher-energy photons to achieve high quantum yield operation at the expense of a large energy spread of electrons. A superlattice photocathode, on the other hand, enables the achievement of high quantum yield and small energy spread (ΔE_{CB}) simultaneously by choosing excitation photon energy (Eg^{SL}) immediately above the step energy of DOS. (2) Larger band-gap energy and smaller electron affinity of the superlattice semiconductor than those of GaAs semiconductor are possible by optimization of semiconductor materials and a superlattice structure. Experimental results of polarized electron sources have suggested that a semiconductor with larger band-gap energy is more suitable for producing a photocathode with high quantum yield [9]. A semiconductor with smaller electron affinity is more suitable photocathode with an NEA surface for maintaining an NEA state for a longer time, because the semiconductor can has larger energy difference between vacuum level and the conduction band minimum when the surface of the semiconductor is activated to the NEA surface [7].

To optimize the superlattice structure for a small energy spread of photoelectrons, we calculated the energy width of the conduction mini band for various AlGaAs-GaAs superlattice structures using the Klonig-Penny-Bastard model [10]. From these results we could observe that AlGaAs-GaAs superlattice structures with Al fraction = 0.29, the well layer thickness below 6 nm and the barrier layer thickness over 6 nm enable energy width of the conduction mini band to be smaller than the thermal energy at room temperature [11].

3. Experimental

In order to confirm small energy spread of photoelectrons from a superlattice photocathode, we decided to develop an AlGaAs-GaAs superlattice semiconductor and measure the dependence of quantum yield on the excitation energy by using the photocathode test system at RIKEN.

3.1. AlGaAs-GaAs semiconductor sample

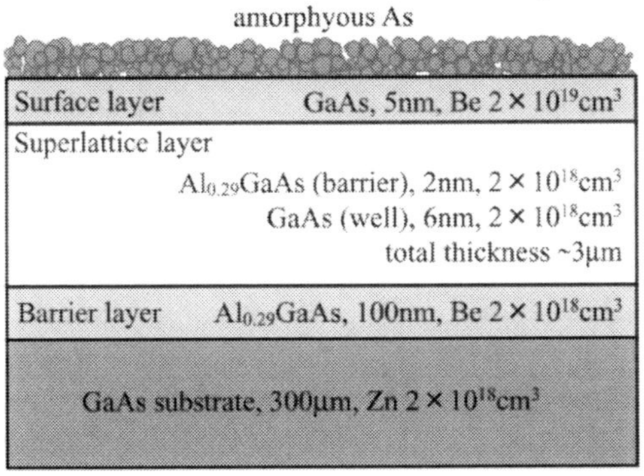

Figure 2. Crystal structure and doping densities of the AlGaAs-GaAs superlattice semiconductor test sample.

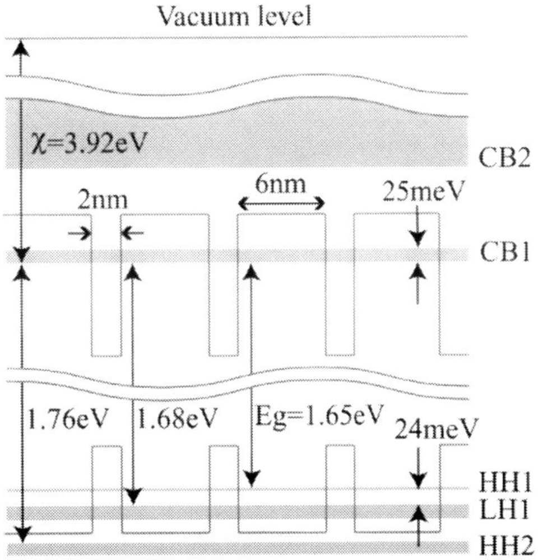

Figure 3. The energy band structure of the AlGaAs-GaAs superlattice semiconductor test sample. The CB, HH, LH are the conduction mini-band, heavy hole mini-band and light hole mini-band. The χ is the electron affinity.

Figure 2 shows the design parameter of the crystal structure of the AlGaAs-GaAs superlattice semiconductor used in this study. A commercially available GaAs wafer with Zn doping of $2\times10^{18}cm^3$ was used a substrate. The thickness of the GaAs well layer is 2nm and the thickness of the AlGaAs barrier layer with Al fraction of 0.29 is 6nm. The 380 pair of the well and the barrier layer with Be doping of $2\times10^{18}cm^3$ is terminated with a heavily doped surface layer of a GaAs (Be $2\times10^{19}cm^3$) to suppress the surface charge limit phenomenon [12]. The superlattice semiconductor is prepared with a protective surface film of amorphous Arsenic, which is removed by annealing at 400 degree in a vacuum. The superlattice semiconductor was fabricated by molecular-beam epitaxy at Venture Business Laboratory of Nagoya University.

Figure 3 shows the energy band structures of the superlattice semiconductor. The energy band structures were obtained from a calculation using the Kronig-Penny-Bastard model. The estimated band gap energy (1.65 eV), which is the energy separation between the bottom of the minimum conduction mini-band (CB1) and the top of the maximum valence mini-band, is larger than that of GaAs (1.43 eV [13]). The electron affinity (χ = 3.92eV) is smaller than that of a GaAs (4.07eV [14]). The energy width of the CB1 is 25meV correspond to the thermal energy at 300K.

3.2. Photocathode test system

The dependence of the quantum yield of the superlattice semiconductor after NEA activation on the excitation energy is measured by using the photocathode test system shown in Figure 4. The system consists of the extremely high vacuum chamber for the NEA-surface activation and a YAG-laser-pumped Ti:Sapphire tuneable wavelength laser.

A base pressure of the extremely high vacuum chamber was 5×10^{-10}Pa. In the extremely high vacuum chamber, small amounts of caesium and oxygen were alternatively supplied to obtain NEA surface after the semiconductor is heat-

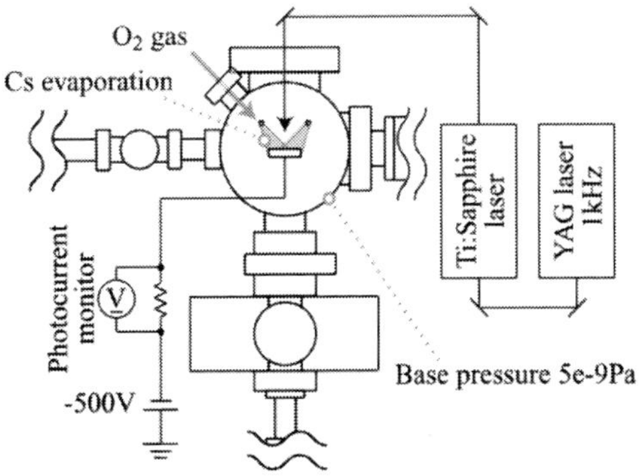

Figure 4. A schematic view of the cathode test system at RIKEN.

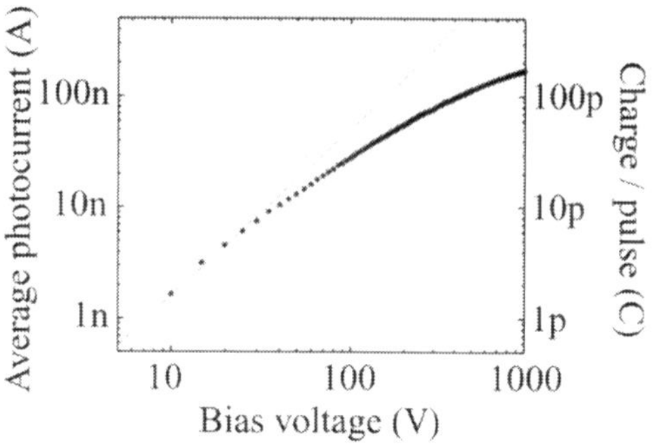

Figure 5. Dependence of the space-charge limited current on the bias voltage. The dot is the extracted photocurrent by using the cathode test system. The dotted line is the child-langmuir's law.

cleaned to remove the protective surface film of amorphous arsenic at 400℃ for 1 h. A commercially available dispenser of the caesium source is on the front of the cathode holder. Oxygen is introduced into the gun chamber from a pure oxygen bottle through a variable leak valve. The quantum yield is monitored by a simultaneous measurement of photocurrent and irradiation laser power. A 633nm He-Ne CW laser is used as the light source to monitor the photocurrent during the NEA activation process, where the semiconductor is biased at -100 V. The YAG-laser-pumped Ti:Sapphire laser is used as a light source with tuneable wavelength for the measurement of the dependence of the quantum yield on the excitation energy, where the semiconductor is biased at -500 V. Extracted average photocurrent during the measurement of the dependence of the quantum yield on the excitation energy is below 5nA which is enough small to avoid serious the space charge limitation effect shown in Figure 5.

4. Result and Discussion

Figure 6 shows the result of quantum yield depend on the excitation energy for the AlGaAs-GaAs superlattice semiconductor after the NEA surface activation. The spectrum has two steps which have the peaks at 1.68 eV and 1.74 eV. It seems that the quantum yield peak around 1.68eV is the spectrum by photoelectrons excited to the CB1 from HH1 and LH1 band, and the quantum yield peak around 1.74eV is the spectrum by photoelectrons excited to the CB1 from the HH2 band. The spectrum around those peaks can be caused by the CB1 with energy width of 25meV, because quantum yield at 1.64eV is higher 4 times than that at 1.67eV and quantum yield at 1.72eV is higher 4 times than that at 1.75eV. The quantum yield at 1.68 eV is 1.1×10^{-4}, the quantum yield at 1.74 eV is 2.8×10^{-4}. These very low quantum yields compared with that of the conventional GaAs photocathode around band gap energy are probably caused by an NEA-surface activation of a poor quality.

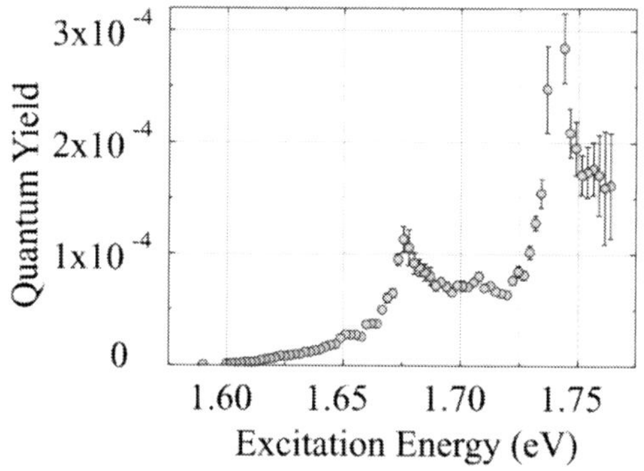

Figure 6. Observed quantum yield depend on the excitation energy for the AlGaAs-GaAs superlattice semiconductor test sample after the NEA surface activation.

5. Conclusions

We considered that a superlattice semiconductor can be more suitable a photocathode for small energy spread of photoelectrons and high quantum yield. We developed the AlGaAs-GaAs photocathode with Al fraction of 0.29, the barrier layer with the thickness of 6nm and the well layer with the thickness of 2nm, and measured the dependence of quantum yield on the excitation energy for the AlGaAs-GaAs superlattice semiconductor test sample after NEA-surface activation. The obtained quantum yield spectrum was a step function of excitation photon energy most likely caused by the quantum confinement effect. Momentum of excited electrons can redistributed during emission through the surface band bending region. If the redistribution could be avoided by a suitable design of the surface region, we conclude that the AlGaAs-GaAs superlattice semiconductor is a suitable NEA-

GaAs photocathode which fulfils the requirements of smaller energy spread of photoelectrons than that of the conventional bulk-GaAs photocathode.

Acknowledgments
This work was partially supported in part by a Grant-in-Aid for Scientific Research (18740253) from the Ministry of Education, Culture, Sports, Science and Technology, and the Kazato Research Encouragement Prize from the Kazato Research Foundation.

References

[1] Omori T, Kurihara Y, Nakanishi T, Aoyagi H, Baba T, Furuya T, Itoga K, Mizuta M, Nakamura S, Takeuchi Y, Tsubata M, and Yoshioka M 1991 *Phys. Rev. Lett.* **67** 3294.
[2] Kurihara Y, Omori T, Nakanishi T, Aoyagi H, Baba T, Itoga K, Mizuta M, Nakamura S, Takeuchi Y, Tsubata M, and Yoshioka M 1992 *Nucl. Instrum. Methods Phys. Res. Sect. A* **313** 393.
[3] Kurihara Y, Omori Takeuchi Y, Yoshioka M, Nakanishi T, Okumi S, Tawada M, Togawa K, Tsubata M, Baba T, Mizuta M, Alley R K, Aoyagi H, Clendenin J E, Frisch J C, Mulhollan G A, Sáez P J, Schultz D C, Tang H and Witte K H 1995 *Japan. J. Appl. Phys.* **34** 355.
[4] SLD Collaboration 1995 Physical Review Letter 74 p2880.
[5] Siggins T, Sinclair S, Bohn C, Bullard D, Douglas D, Grippo A, Gubeli J, Krafft G A and Yunn B 2001 *Nucl. Instrum. Methods Phys. Res. Sect.* **A 475** 549.
[6] Grames J, Adderley P, Brittian J, Charles D, Clark J, Hansknechta J, Poelker M, Stutzman M, Surles-Law K 2005 *Proc. of Particle Accelerator Conf.* 2875.
[7] Nishitani T, Hajima R, Iijima H, Nagai R, Sawamura M, Kikuzawa N, Nishimori N, Minehara E, Tabuchi M, Noritake Y, Hayashitani H, and Takeda Y 2006 *Proc. Int. Free Electron Laser Conf.* 319.
[8] Nishitani T, Nakanishi T, Yamamoto M, Okumi S, Furuta F, Miyamoto M, Kuwahara M, Yamamoto N, Naniwa K, Watanabe O, Takeda Y, Kobayakawa H, Takashima Y, Horinaka H, Matsuyama T, Togawa K, Saka T, Tawada M, Omori T, Kurihara Y, Yoshioka M, Kato K, and Baba T 2005 *J. Appl. Phys.* **97** 094907.
[9] Nakanishi T, Okumi S, Togawa K, Takahashi C, Suzuki C, Furuta F, Ida T, Wada K, Omori T, Kurihara Y, Tawada M, Yoshioka M, Horinaka H, Wada K, Matsuyama T, Baba T, and Mizuta M 1998 *American Institute of Physics Conf. Proc.* **421** 300
[10] Bastard G, 1981 *Phys. Rev.* **B 24** 5693
[11] Nishitani T, Tabuchi M, Takeda Y, Suzuki Y, Motoki K, Meguro T 2009 *Japan. J. Appl. Phys.* **48** 06FF02
[12] Subashiev A V, Mamaev Y A and Yashin Y P, Clendenin J E 1998 *SLAC-PUB* 8035
[13] Lautenschlager P, Garriga M, Logothetidis S and M Cardona 1987 *Phys. Rev.* **B 35** 17 9174
[14] Gobeli G W and Allen F G 1965 *Phys. Rev.* **137** 1A A245

Elimination of local thickness modulation in GaAs/GaAsP strained superlattices for high spin-polarization photocathodes

Xiuguang Jin[1a)], Yuya Maeda[1], Toshio Sasaki[2], Shigeo Arai[2], Shingo Fuchi[1], Toru Ujihara[1], Yoshikazu Takeda[1]

[1] Department of Crystalline Materials Science, Graduate School of Engineering, Nagoya University, Furo-cho, Chikusa-ku, Nagoya 464-8603, Japan

[2] High Voltage Electron Microscope Laboratory, Eco Topia Science Institute, Nagoya University, Furo-cho, Chikusa-ku, Nagoya 464-8603, Japan

E-mail: jin@mercury.numse.nagoya-u.ac.jp

Abstract. We successfully developed a transmission-type photocathode, and a high spin-polarization (90%) with a super-high brightness (1.3×10^7 $Acm^{-2}sr^{-1}$) of electron beam was achieved. In this research, the elimination of thickness modulation of GaAs/GaAsP superlattice by introduction of a GaAs inter-layer on a GaP substrate is the key point to realize the high spin-polarization. The thickness modulation of the superlattice is related with the surface roughness of the buffer layer. The compressive strain introduced in the GaAsP buffer layer on the GaP substrate causes hillock formation, where several degree off-angle surfaces are formed. The GaAs inter-layer deposited on the GaP substrate introduced a tensile strain in the GaAsP buffer layer instead of the compressive strain and relatively smooth GaAsP buffer layer was achieved. The smooth GaAsP buffer layer was attributed to the periodic GaAs/GaAsP superlattice layer growth.

1. Introduction

Highly spin-polarized electron sources are intensively developed for applications in high-energy physics, atomic physics, and studies of thin film and surface magnetism [1]. A strained superlattice structures composed of a GaAs-related semiconductors are known to be most effective as photocathode of the spin-polarized electron beam [2, 3]. In these structures, the heavy hole and light hole mini-bands are largely split due to both the elastic strain and the quantum confinement effect. So the highly spin-polarized electrons are selectively excited by circularly polarized photons only from the heavy hole mini-band to the conduction mini-band, and are then emitted from a negative electron affinity surface. Photocathodes based on a GaAs/GaAsP strained superlattice [4, 5] and an AlInGaAs/AlGaAs strained superlattice [6] provide high spin-polarizations over 90%.

In our previous study [5, 7] we have newly developed a super-high-brightness transmission-type photocathode based on the GaAs/GaAsP strained superlattice. In this novel system, a super-high brightness electron beam was achieved by illuminating a small area on the backside of the photocathode with a pump laser light that was well-focused by a short-focal-length lens. To realize this system, a GaAs/GaAsP strained superlattice was fabricated on a GaP substrate (bandgap energy: 2.26 eV), which is transparent to the pump-laser beam (beam energy: 1.44–1.77 eV), instead of a GaAs (bandgap energy: 1.42 eV) substrate. High spin-polarization (90%) and a super-high brightness (1.3×10^7 $A\cdot cm^{-2}\cdot sr^{-1}$) electron beam were reproducibly achieved with this transmission-type photocathode. The key technology to achieve the high spin-polarization in this photocathode is the

[a)]Electronic mail: jin@mercury.numse.nagoya-u.ac.jp

introduction of a 500 nm GaAs inter-layer deposited on the GaP substrate. The spin-polarization of the photocathode without the GaAs inter-layer was approximately 60%.

In this study, the superlattice structure on a GaAs substrate, GaP substrate, and a GaP substrate with the GaAs inter-layer were investigated by cross-sectional transmission electron microscopy (TEM) and atomic force microscopy (AFM). The effect of the GaAs inter-layer on spin-polarization is discussed.

2. Experimental

The photocathodes were grown using a low-pressure metalorganic vapor phase epitaxy system with a vertical cold-wall quartz reactor. Triethylgallium (TEG), tertialybutylphosphine (TBP), and tertialybutylarsine (TBAs) were used as source materials. The substrates were Zn-doped p-type and (001)-oriented GaAs and GaP. All samples were grown at 660°C under a reactor pressure of 76 Torr. The photocathode structures are shown in Figure 1. Following the growth of 1 or 2 μm of the GaAsP buffer layer on the GaAs substrate, the GaP substrate, or the GaP substrate with the 500 nm GaAs inter-layer, a 12-pair GaAs/GaAsP strained superlattice was grown with a Zn doping to a concentration of 1.5×10^{18} cm^{-3}. The thickness of the GaAs and GaAsP was 4 nm. Subsequently, each superlattice structure was coated with a highly-doped 5 nm GaAs layer with a Zn dopant concentration of $6\times10^{19}cm^{-3}$. The spin-polarizations of the photocathodes on the GaAs substrate and the GaP substrate with the GaAs inter layer were over 90%, while that on the GaP substrate was only 60%. TEM observation was performed at the electron energy of 200 keV using a Hitachi H-800 electron microscopy. The surface morphology was observed using AFM (NanoScope III).

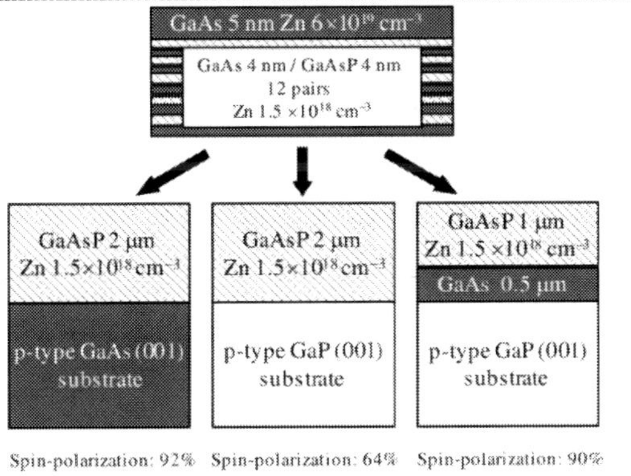

Figure 1. Details of the photocathode structures. GaAs/GaAsP superlattices were grown on buffer layers with different substrates: GaAs, GaP, and GaP substrate with GaAs inter-layer.

3. Results and discussion

3.1 TEM image of GaAsP buffer layers and GaAs/GaAsP superlattice layers

Figure 2 shows the cross-sectional TEM images of the GaAsP buffer layers and superlattice layers on the various substrates. A stacking fault extended from the GaAsP buffer layer into the GaAs/GaAsP superlattice layer, as shown in figure 2 (a). In other areas, the period of the superlattice layers are well controlled. Figure 2 (b) shows the TEM image of the superlattice and the GaAsP buffer layer on the GaP substrate. The remarkable feature in this sample is the bump structures. The bump structures take place on the vicinal GaAsP buffer layer in which the surface inclined by about 6° from (001) plane. In the bump structure, the layer thickness in the superlattice is modulated. The thickness modulation increases with the increasing number of the layer pairs (gray/black) and is especially enhanced after the fifth pair. A thicker GaAsP barrier-layer (gray) is grown on a thinner GaAs well-layer (black), while a thin GaAsP barrier-layer is grown on a thicker GaAs well-layer. On the other hand, the periodic superlattice layers are grown on the flat GaAsP buffer layer. The TEM images of the superlattice and the GaAsP buffer layer on the GaP substrate with the GaAs inter-layer are shown in Figure 2 (c). The surface of the superlattice is smooth with no bump structure. The image is similar to that of the superlattice on the GaAs substrate, i.e., the superlattice layers are periodic and stacking faults penetrate from the buffer layer into the superlattice.

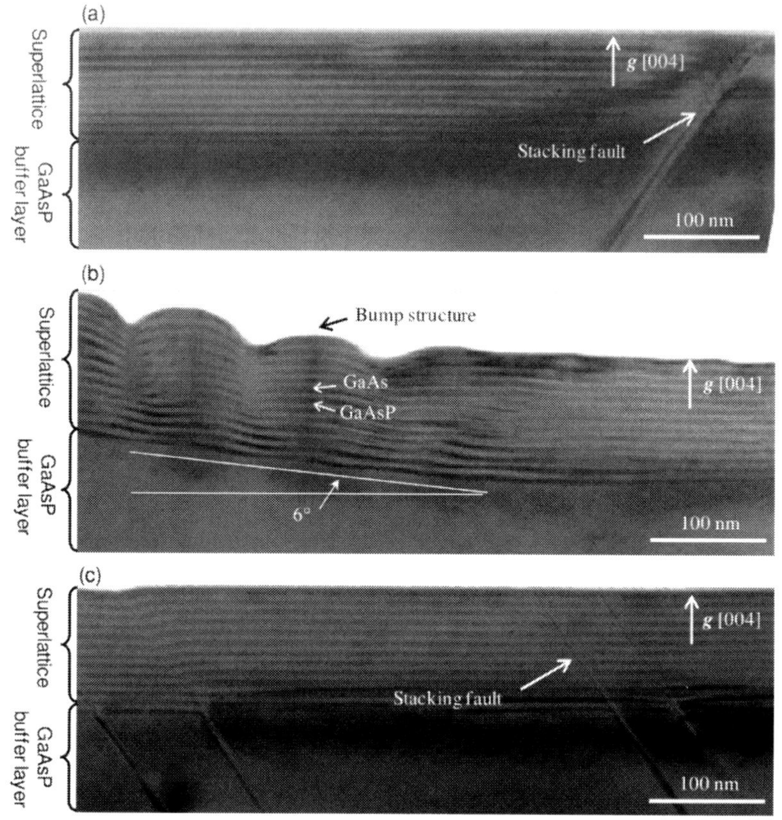

Figure 2. Cross-sectional TEM images of GaAs/GaAsP strained superlattice layer and GaAsP buffer layer on (a) GaAs, (b) GaP, and (c) a GaP substrate with GaAs inter-layer. In (b), bump structures observed in the suprelattice layer.

3.2 Surface morphology of GaAsP buffer layer on various substrates

The surface morphology of GaAsP buffer layers were evaluated using AFM and are shown in Figure 3 (a)~(c). In the GaAsP layer on the GaAs substrate, some cracks along [110] and [1-10] directions were observed as shown in Figure 3 (a). Except the cracks areas, the surface of GaAsP layer shows relatively flat. The surface of the GaAsP layer on the GaP substrate has some anisotropic hillocks as shown in Figure 3 (b). The size of the hillocks structure is approximately 100 nm height and 2 μm wide. The surrounding of the hillocks show vicinal surface and the vicinal surfaces inclined by about 5° from (001) plane. Figure 3 (c) shows the surface of the GaAsP layer on the GaP substrate with GaAs inter-layer. The surface roughness was much better controlled by introduction of the GaAs inter-layer on the GaP substrate.

3.3 Discussion

In the superlattice on the GaAs substrate and the GaP substrate with the GaAs inter-layer, the stacking fault is the main defect and the density is in the range of 10–15 per 10μm. In spintronics research, spin-flip by defect has been reported. In spin-polarized light-emitting diodes (spin-LED)[8], spin-polarized electrons are injected from ZnMnSe into a GaAs/AlGaAs quantum well (QW), where radiative recombination of the carriers results in the emission of a circularly polarized light. The spin-

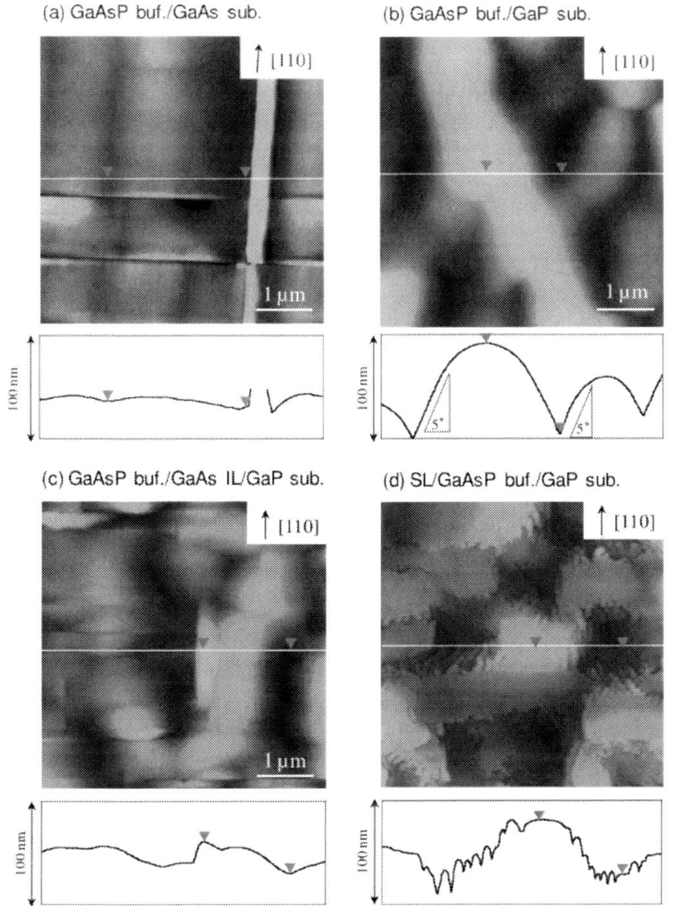

Figure 3. AFM images of GaAsP buffer layer on (a) GaAs, (b) GaP, (c) a GaP substrate with GaAs inter-layer and (d) GaAs/GaAsP superlattice and GaAsP buffer layer on GaP substrate. In (d), bump structure surround the hillock structures.

polarization of the injected electron correlates inversely with the stacking fault density at the ZnMnSe/AlGaAs interface. According to the experiments, spin-polarization of ~90% can be realized even with the stacking fault density of 10–15 per 10μm. The value of the spin-polarization is almost the same as those in the present study. Therefore, we believe that the effect of the stacking fault (10–15 per 10 μm) on the spin-polarization is not serious.

In the strained superlattice, the heavy hole mini-band and the light hole mini-band are split due to both the quantum confinement and the elastic strain effect. Adjusting the excitation photon energy to the energy gap between the heavy hole mini-band and the conduction mini-band, the spin-polarization of electron beam reaches a maximum value. The energy band gap greatly depends on the thicknesses of the superlattice layer. According to the calculation of the band structure, we found that the energy gap in the thickness modulation area is smaller than that in the periodic area. The difference of the energy gap in both areas is larger than the valence band split. Therefore, it is possible that electrons of both up-spin and down-spin are excited from the heavy hole mini-band in the periodic areas and from the light hole mini-band in the thickness modulated areas by the same energy of the exited photons. We consider the spin-polarization degradation in the superlattice on the GaP substrate is related with the thickness modulation.

In the crystal growth, the thickness modulation of the superlattice on the vicinal surface was reported by several groups [9-11] and it`s mechanism was attributed to the elastic relaxation [10]. The lattice constant of GaAsP is larger than that of GaP, and smaller than that of GaAs. Therefore, a compressive strain introduces in the GaAsP buffer layer grown on the GaP substrate. The compressive strain usually causes hillock formation, where several degree off-angle surfaces are formed. As shown in Figure 3 (d), the bump structures were formed on the surrounding of the hillocks. The GaAs inter-layer deposited on the GaP substrate introduced a tensile strain in the GaAsP buffer layer instead of the compressive strain and relatively smooth GaAsP buffer layer was achieved. The smooth GaAsP buffer layer was attributes to the periodic GaAs/GaAsP superlattice layer growth. As a result, high spin-polarization of 90 % was achieved by introduction of the GaAs inter-layer on the GaP substrate.

4. Conclusion

Superlattice structure on a GaAs substrate, GaP substrate, and a GaP substrate with a GaAs inter-layer, as spin-polarized photocathodes, were investigated by TEM and AFM. In the superlattice on the GaP substrate, the thickness modulation of the superlattice is the main reason to reduce the spin-polarization. The thickness modulation of the superlattice related with the roughness of the buffer layer. The introduction of the GaAs inter-layer on the GaP substrate controlled the surface roughness of the GaAsP buffer layer and was attributed to the periodic superlattice growth.

Acknoledgements

This work was partially supported by the Technology Development Program for Advanced Measurement and Analysis of the Japan Science and Technology Agency (JST) and Research for Promoting Technology Seeds. One of the authors (X.G. Jin) is supported by the Japan Society for the Promotion of Science (JSPS) Research fellowship for young scientists.

Reference

[1] Yamamoto N, Nakanishi T, Mano A, Nakagawa Y, Okumi S, Yamamoto M, Konomi T, Jin X G, Ujihara T, Takeda Y, Ohshima T, Saka T, Kato T, Horinaka T, Yasue T, Koshikawa T and Kuwahara M 2008 J. Appl. Phys. **103** 064905

[2] Omori T, Kurihara Y, Takeuchi Y, Yoshioka M, Nakanishi T, Okumi S, Tsubata M, Tawada M, Tanimoto Y, Takahashi C, Baba T, and Mizuta M 1994 Jpn. J. Appl. Phys. **33** 5676

[3] Saka T, Kato T, Nakanishi T, Okumi S, Togava K, Horinaka H, Matsuyama T, and Baba T 2000 Surf. Sci. **454** 1042

[4] Nishitani T, Nakanishi T, Yamamoto M, Okumi S, Furuta F, Miyamoto M, Kuwahara M, Yamamoto N, Naniwa K, Horinaka H. Matsuyama T, Togawa K, Saka T, Tawada M, Omori T, Kurihara Y, Yoshioka M, Kato K, and Baba T 2005 J. Appl. Phys. **97** 094907

[5] Jin X G, Yamamoto N, Nakagawa Y, Mano A, Kato T, Tanioku M, Ujihara T, Takeda Y, Okumi S, Yamamoto M, Nakanishi T, Saka T, Horinaka H, Kato T, Yasue T, and Koshikawa T 2008 Appl. Phys. Express **1** 45002

[6] Mamaev Yu A, Gerchikov L G, Yashin Yu P, Vasiliev D A, Kuzmichev V V, Ustinov V M, Zhukov A E, Mikhrin V S, and Vasiliev A P 2008 Appl. Phys. Lett. **93** 081114

[7] Jin X G, Maeda Y, Saka T, Tanioku M, Fuchi S, Ujihara T, Takeda Y, Yamamoto N, Nakagawa Y, Mano A, Okumi S, Yamamoto M, Nakanishi T, Horinaka H, Kato T, Yasue T, and Koshikawa T 2008 J. Cryst. Growth **310** 5039

[8] Stroud R M, Hanbicki A T, Park Y D, Kioseoglou G, Petukhov A G, and Jonker B T 2002 Phys. Rev. Lett. **89** 166602

[9] Cox H M, Aspnes D E, Allen S J, Bastos P, Hwang D M, Mahajan S, Shahid M A and Morais P C 1990 Appl. Phys. Lett. **57** 611

[10] Ponchet A, Rocher A, Emery J -Y, Starck C and Goldstein L 1993 J. Appl. Phys. **74** 3778

[11] Chen Y, Chiu T H, Zucher J E and Chu S N G 1993 Appl. Phys. Lett. **62** 3201

A study on XAFS analysis of Cs/GaAs NEA surface

Atsushi Era[a], Masao Tabuchi[a], Tomohiro Nishitani[b], and Yoshikazu Takeda[a]

[a] *Nagoya University, Furo-cho, Chikusa-ku, Nagoya 464-8603, Japan*

[b] *RIKEN, 2-1 Hirosawa, Wako, Saitama 351-0198, Japan*

E-mail: era@mercury.numse.nagoya-u.ac.jp

Abstract. Although NEA (negative electron affinity) surface has been attracting great interest, atomic level structure of the NEA surface has not been elucidated. In this work, Cs/GaAs NEA surfaces were analyzed by fluorescence XAFS (x-ray adsorption fine structure) method. Cs-L_{III} XAFS measurement was conducted on the Cs/GaAs surfaces obtained by Yo-Yo method. By analyzing the local structure around Cs from the XAFS spectra, it was shown that Cs-O-As (or Ga) or O-Cs-As (or Ga) structure might be formed on Cs/GaAs surface. The results suggested that single atomic layer of Cs was adsorbed on GaAs surface intermediated by O on the Cs/GaAs NEA surface.

1. Introduction

NEA (negative electron affinity) surface has been attracting great attention in recent years, because of its availability for a highly functional electron sources. A GaAs surface covered by Cs (Cs/GaAs) is one of the basic structures of the NEA surface. In the Cs/GaAs NEA system, when an electron is excited in GaAs by irradiation of light, the excited electron is easily emitted from the surface. Therefore electron emission can be controlled by irradiation of light and of which energy distribution can be controlled by the energy of the irradiating light. Here, the NEA-photocathode can be a very low-emittance electron source, and can be operated as very short pulse or high repetition rate electron source by controlling the pulse width and repetition rate of the excitation light. The NEA-photocathode is expected to be utilized in the next generation accelerators and in the highly functional electron microscopes, e.g., time resolved electron microscope.

In order to utilize the NEA surface as a production level photo-cathode, the life time should be longer and the quantum efficiency (QE) should be higher. To achieve the longer life time and the higher QE, how the NEA is realized at the surface must be well understood, and for that, the structure of the surface formed by Cs adsorption on GaAs must be well recognized. Several experimental studies using low-energy electron diffraction, scanning tunneling microscopy and photoemission experiment have been performed to investigate the NEA surface [1-4]. Nevertheless, the state of the NEA surface is not still understood well. A few structure models are proposed [5, 6]. However, the actual NEA surface structure is still not clarified.

In this study, we investigated the NEA surface by using X-ray absorption fine structure (XAFS) analysis. The XAFS measurement is a kind of photo-absorption measurement in the X-ray level energy range. When the X-ray absorption measurement is conducted in a wide energy range, jumps are observed on the absorption coefficient spectrum that is caused by excitation of core-level electrons of atoms in target material. When we measure the spectrum precisely after a jump edge, fine structure is observed, which is the XAFS spectrum. The fine structure reflects information on the atomic level local structure around the X-ray-absorbing atom that causes the jump on the spectrum. Therefore, by measuring and analyzing the XAFS spectra above the jump edge caused by Cs, we can discuss on the local structure formed with Cs on the GaAs surface.

2. Experiments

Be-doped p-type GaAs layers were grown on a GaAs substrate by molecular beam epitaxy (MBE) method, and the surface was capped by amorphous As in the same ultra-high vacuum MBE chamber. Then, the sample was transferred from the MBE chamber to an NEA activation chamber through the air. The amorphous As layer prevents the GaAs surface from pollutions during the transfer through the air. The amorphous As layer was removed by heating the sample in the NEA activation chamber. Then, the activated NEA surface was prepared by supplying Cs and oxygen alternatively monitoring the photo-current. The bias voltage to monitor the photo-current was −100V. The alternative supplying process to obtain the NEA surface is so called Yo-Yo method [7]. The Cs and oxygen on the NEA activated surface is easily removed by heat cleaning at a controlled temperature. After the heat cleaning, the surface is considered to be equivalent to the fresh GaAs surface before the activation process. In this work, the activation process and the heat cleaning were repeated several times.

The NEA activation chamber was specially designed for the in-situ XAFS measurement. The chamber was installed on the XAFS measurement line using synchrotron radiation at BL9A of Photon Factory in Tsukuba, Japan. A monochromatized X-ray was irradiated on the sample in the NEA activation chamber through a Be window. The Fluorescence X-ray from the sample was detected by 19 elements Ge-Solid-State-Detector (SSD). The XAFS measurement was conducted for the Cs-L_{III} absorption edge (5.011 keV) by sweeping the incident X-ray energy from 4.8 to 5.4 keV. The electric field of the incident X-ray was polarized normal to the surface. Under the condition, measured XAFS spectra strongly reflected the local structure normal to the surface and were not sensitive to the in-plane structure. The base pressure of the NEA activation chamber was kept at about 3×10^{-8} Pa during the XAFS measurement by using ion pump and non-evaporable getter (NEG) pump. As described above, the NEA activation process was conducted several times, and the XAFS measurement was conducted after each activation process. We spent about 15 hours to obtain one XAFS spectrum. QE and Cs-L fluorescence X-ray yield were checked several times during each XAFS measurement. The QE of the NEA surface was different after each activation process. In this paper, we discuss on the NEA surface of the highest and the lowest quantum efficiency.

3. Results of XAFS measurements

Figure 1 shows the change of the measured QE checked using 633nm He-Ne laser during the XAFS measurement. The initial QE of the highest and the lowest QE surfaces was about 3 and 0.05 %, respectively. As shown in Fig. 1, the QE was decreased as usual depending on time. For the highest QE surface, it decayed by about two orders of magnitude and for the lowest one it decayed by about four orders of magnitude during the XAFS measurement. Figure 2 shows the fluorescence X-ray spectra. The solid lines show spectra measured just after the activation process, and the dotted lines are spectra measured after about 15 hours. The peaks observed at about 4.2 keV are Cs-L fluorescence, and the peaks observed at about 5.0 keV are due to the elastically scattered incident X-ray. As shown in Fig. 2, almost no change was observed on the intensity of Cs-L fluorescence peak depending on time, while the quantum efficiency decayed. This result indicated that the decreasing of the QE with time was not caused by desorption of Cs from the surface during the XAFS measurement. Therefore, there must be some structural change on the surface relating to the change of the QE. However, in this work, we assumed that the structural change throughout the XAFS measurement time (for about 15 hours) was smaller than the difference of the structures between the highest and the lowest ones. Since we need the 15 hours to obtain one XAFS spectrum and could not resolve the

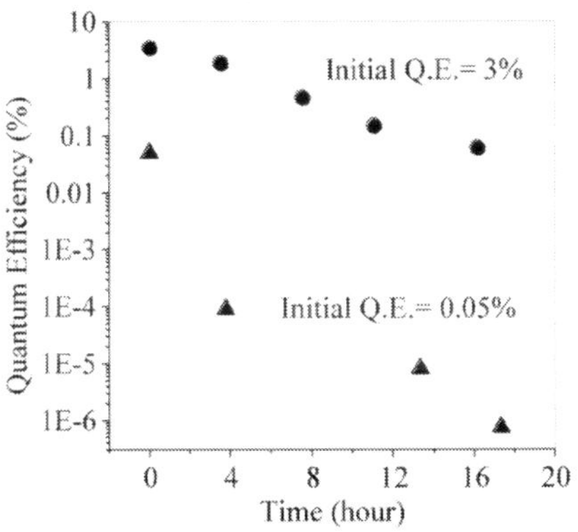

Fig. 1. Change of the QE during the XAFS measurement. Circles show the highest QE surface. Triangles show the lowest QE surface.

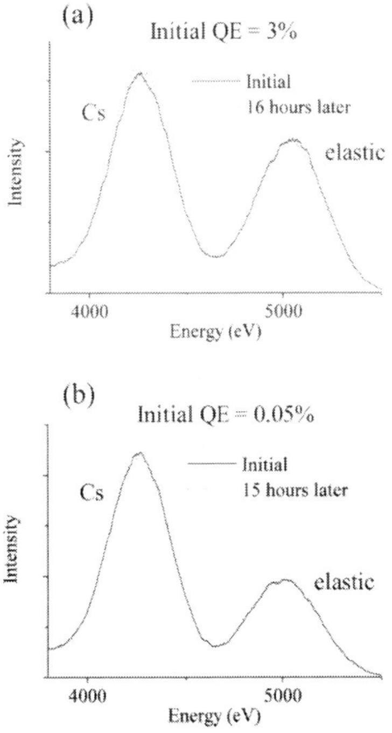

Fig. 2. Fluorescence X-ray yield spectra measured during XAFS measurement for the highest (a) and the lowest (b) QE surface.

structural change faster than that. Figure 3 shows the measured XAFS spectra for the highest and the lowest QE surfaces, and normalized XAFS oscillations are shown in Fig. 4. As shown in Fig. 3 and 4, the quality of the spectra was not so good. It means the amounts of Cs atom on the GaAs surface were very small. Although we could not decide it quantitatively, the amount should be about 1 or 2 atomic layers based on our experience. In spite of the low quality of the spectra, it is found that these two spectra showed different features, which indicated that the structures formed around Cs atom were different for the highest and the lowest QE surfaces. Figure 5 shows the Fourier transformed XAFS spectra comparing with some theoretical calculated ones by using FEFF code [8], which is generally used to calculate XAFS spectra theoretically. The Fourier transformed XAFS spectra are known to reflect the radial distributions of atoms around the X-ray absorbing atom (in this case, it is Cs). Spectra (a) and (b) are measured ones for the highest and the lowest QE surfaces. Other spectra are calculated ones for the model structures showed on the right side in Fig. 5. In the calculations, the distances between atoms were adjusted to represent the two peaks at about 2 and 4 Å as in the measured spectra. It was obvious that the spectra (a) and (b) are different. It indicates again Cs formed different structure on the highest and the lowest QE surfaces. By comparing the measured spectra to the calculated ones, it was found that all the four models did not give a spectrum similar to that of the lowest QE surface. Therefore, we could not discuss on the precise structure of the lowest QE surface now. On the other hand, for the highest QE surface, the Cs-O-As (Model-1) and O-Cs-As (Model-2) models gave the spectra similar to the measured one of the best QE surface. It should be noted that As and Ga was very difficult to distinguish in the XAFS measurement. Therefore, As atoms in the models might be Ga atoms.

4. First principle calculations

Although, it was difficult to analyze the XAFS spectra quantitatively since the quality of the spectra was not good, we tried further consideration on the Model-1 and -2 to find which one is more probable structure. First principle calculations were conducted for the Cs-O-As and O-Cs-As models to find the Cs-O and Cs-As distances that give the minimum energy. In the calculation the positions of Ga and As atoms of GaAs crystal were fixed. Changing the positions of Cs and O, the total energy

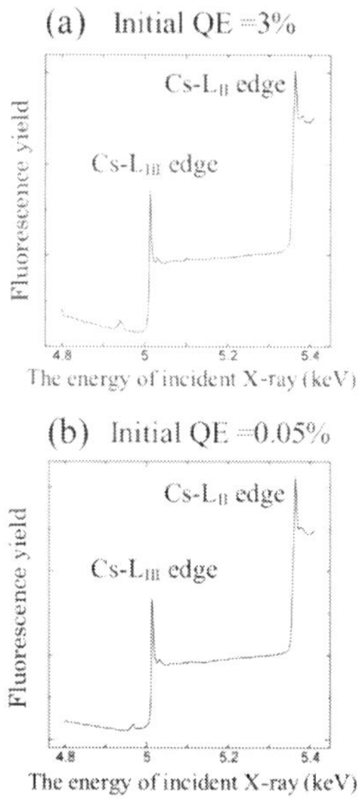

Fig. 3. Measured X-ray absorption spectra for the highest (a) and the lowest (b) QE surfaces.

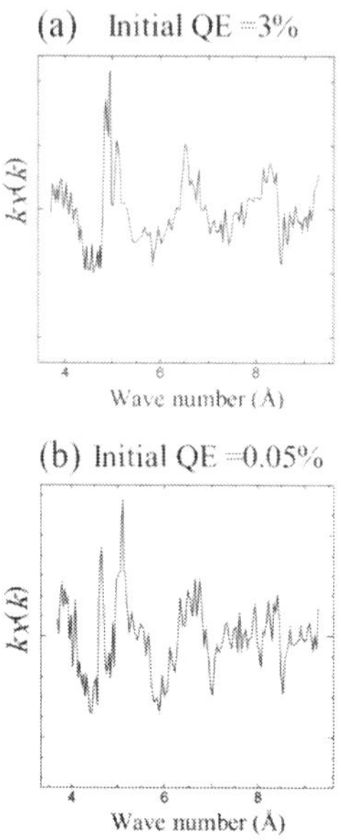

Fig. 4. Normalized XAFS oscillations for the highest (a) and the lowest (b) QE surfaces.

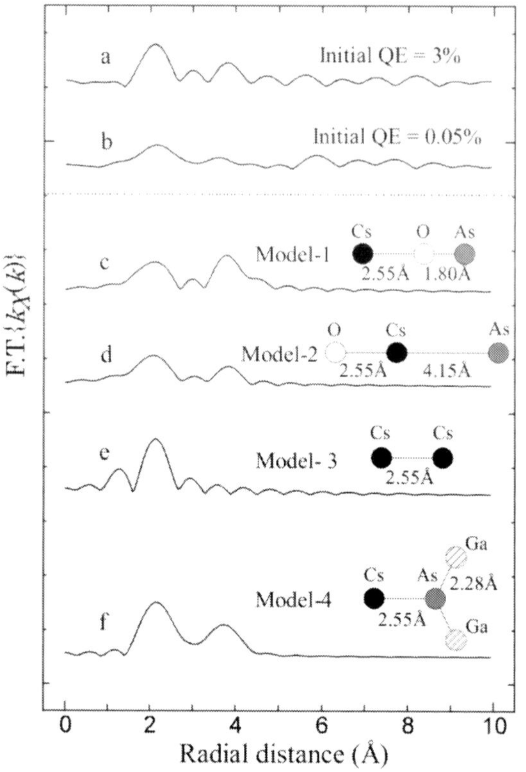

Fig. 5. Fourier transformed XAFS spectra, which reflected the radial distributions of atoms around Cs. (a) and (b) are measured ones for the highest and the lowest QE samples. (c), (d), (e), and (f) are theoretically calculated ones.

was calculated repeatedly. Then the Cs-O and Cs-As distances which give the minimum energy were decided. Figure 6 shows the relationship between the Cs-O and Cs-As distances. As shown in Fig. 6, when we assumed Model-1, the measured and calculated relationships between the Cs-O and Cs-As were very similar. On the other hand, with the Model-2 the measured and calculated results did not agree with each other. Therefore, Cs-O-As structure was more probable than O-Cs-As structure. As mentioned before, the As and Ga atoms were very difficult to distinguish from the XAFS measurement, and the first principal calculations also gave almost the same results for the model with Ga as with As. Therefore, we should conclude that either Cs-O-As or Cs-O-Ga structure might form on the Cs/GaAs surface when the QE was good.

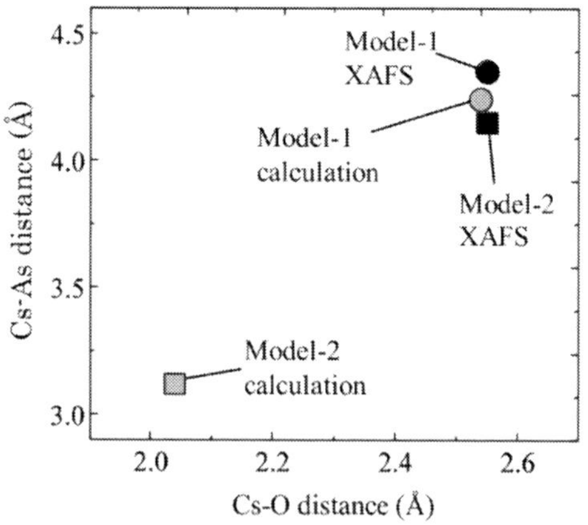

Figure 6. Relationship between the Cs-O and Cs-As distances which are obtained from XAFS analysis and first principle calculation.

5. Summary

XAFS measurements were conducted for Cs/GaAs NEA surfaces. The fluorescence yield measurement during the XAFS measurement showed that the decrease of Q.E. depending on time was not caused by desorption of Cs from surface. The features of the XAFS spectra were different on the highest and the lowest QE surfaces, which indicated that different structure were formed on Cs/GaAs surface with the highest and the lowest QE. The discussion on the precise structure formed on the surface was difficult since the quality of the measured XAFS spectra were not so good. By combining results from XAFS measurement and the first principle method calculation, we found a probable structure formed on the highest QE surface can be Cs-O-As or Ga structure. The results suggested that single atomic layer of Cs was adsorbed on GaAs surface intermediated by O on the Cs/GaAs NEA surface.

References

[1] Yamada T, Fujii J, Mizoguchi T 2001 *Surf. Sci.* **479** 33
[2] Tereshchenko Q E, Voronin V C, Scheibler H E, Alperovich V L and Terekhov A S 2002 *Surf. Sci.* **507-510** 51
[3] Tereshchenko O E, Paget D, Chiaradia P, Wiame F and Taleb-Ibrahimi A 2010 *Phys. Rev. B* **81** 035304
[4] Fouquet P and Witte G 2001 *Appl. Surf. Sci.* **180** 286
[5] Su C Y, Spicer W E and Lindau I 1983 *J. Appl. Phys.* **54** 1413
[6] Sakai J, Mizutani G and Ushioda S 1993 *Surf. Sci.* **283** 217
[7] Togawa K, Nakanishi T, Baba T, Furuta F, Horinaka H, Ida T, Kurihara Y, Matsumoto H, Matsuyama T, Mizuta M, Okumi S, Omori T, Suzuki C, Takeuchi Y, Wada K, Wada K and Yoshioka M 1998 *Nucl. Instr. and Meth. in Phys. Res. A* **414** 431
[8] Rehr J J, Albers R C and Zabinsky S I 1992 *Phys. Rev. Lett.* **69** 3397

Spin polarized electron transport and partial localization of photoelectrons in highly doped photocathodes

L G Gerchikov[1,3], K Aulenbacher[2], Yu A Mamaev[1], E J Riehn[2] and Yu P Yashin[1]

[1]St. Petersburg State Polytechnic University, St. Petersburg, Russia
[2]Institute of Nuclear Physics, Mainz University, Mainz, Germany

E-mail: lgerchikov@rambler.ru

Abstract. The results of experimental and theoretical studies of spin polarized electron transport in semiconductor SL used for photoemitter application are presented. The experimental study is based on the time resolved measurements of electron emission from the cathode after its photoexcitation by fs laser pulse. The response and spin relaxation times have been determined by means of measured time dependent intensity and polarization of electron emission. We also performed theoretical calculations of photocathode pulse response and compared the obtained results with experimental data. Our analysis testifies the presence of partial electron localization in SL. Electron capture by localized states leads to unavoidable losses of quantum efficiency and polarization.

1. Introduction

At the present time, strained semiconductor superlattices (SL) are known as most effective basis of highly polarized electrons sources [1]. Such photoemitters combine the advantages of conventional polarized electrons sources based on a strained semiconductor layer with additional possibilities for band structure engineering of photocathodes working layers based on strained SL. The major goal of SL development for photoemitter applications in the past decade was the achievement of electron polarization more that 90%. For this purpose several types of SLs with highest possible valence band splitting had been developed, namely the SLs with strained quantum wells (QW) [2,3] and with strained barriers [4]. Valence band splitting in SL, i.e. the energy splitting between the upper heavy and light hole minibands, is formed due to the combination of two effects: the strain deformation of SL layers and the quantum confinement. The aim of the optimal SL design is to provide a valence band splitting of more than 70 – 80 meV together with good transport properties and high structural quality.

However, in spite of significant progress in electron polarization (P) the quantum efficiency (QE) at the polarization maximum of the developed photoemitters is rather small. The best combination of P = 92% and QE = 0.85% has been achieved for AlInGaAs/AlGaAs SL with strained QWs [3]. Thus the further progress of SL based photoemitters is shifted towards the developing of highly effective polarized electron sources in order to meet modern requirements of high energy physics [5]. The developing of photocathode structures with distributed Bragg reflector (DBR) is an important step in this direction. One of the most fundamental reasons limiting the QE is the fact that

[3] Corresponding author.

the high polarization of electron emission is achieved at the expense of QE. Indeed, the maximum spin polarization of photoelectrons takes place at the photoabsorption threshold where the photoabsorption coefficient is rather small. Strained SL can not be made too thick due to the possible strain relaxation resulting in structural defects, smaller residual strain and lower polarization. Thus the thickness of the working layer is smaller by an order of magnitude than the photoabsorption length and most of the light intensity is lost in the photocathode substrate. To overcome this problem the photocathode structure with DBR at the back side of the photocathode has been proposed [6-8].

Another way to increase QE is to enlarge the working layer as much as possible and to minimize the losses of photoelectrons prior to their emission into the vacuum. However this traditional approach does not work in the case of SL based cathodes and this fact indicates that the spin polarized electron transport in SLs is still not understood as well as for the conventional photoemitters [9, 10]. The QE of conventional cathodes based on a GaAs working layer gradually increases with the growth of working layer thickness until it exceeds the absorption and/or diffusion length. On the contrary our measurements of QE at polarization maximum for a series of SL based cathodes with a different number of SL periods show that the QE of the cathodes with thick SL containing more that ten periods is not larger than the QE of the cathodes with a four or six period SL. It means that electron transport in thick SLs is suppressed and useful SL length and hence the QE is strongly limited.

The aim of the present work is to study the spin polarized electron transport in SLs in order to figure out the reasons that are responsible for the suppression of electron transport and QE losses. Another aim of our study is to determine the polarization losses that occur on the stage of electron transport from SL to band bending region (BBR). It will be helpful to clarify whether the polarization losses can be reduced even more and the achieved level of electron polarization can be exceeded.

Our experimental method is based on time resolved measurements of electron emission from the cathode after its photoexcitation by a femtosecond (fs) laser pulse. This method has been developed by K. Aulenbacher et al [9] to study the spin polarized electron transport in conventional photocathodes. The observed time dependent intensity of electron emission determines the cathode response time while the time resolved measurement of electron polarization gives the spin relaxation time. We also developed a theoretical description of electron transport in SLs based on the time dependent kinetic equation and calculated the photocathode pulse response. The detailed comparison of the obtained results with the experimental data reveals a significant discrepancy between the observed electron emission and the pulse response of an ideal SL. We interpret this fact as a presence of partial electron localization in SL. Such localization which drastically changes the character of vertical electron transport in SLs has been already observed in heavily doped GaAs/AlAs SLs [11]. The high doping level is used within the photocathode structures in order to form a narrow BBR needed to minimize electron and polarization losses during electron emission into the vacuum. Besides the fluctuations of impurity potential and/or the SL's layer composition the structural defects formed due to the strain relaxation in SLs can be also considered as a source of electron localization.

We developed a simple kinetic model to describe the electron transport in a SL in the presence of partial electron localization. It takes into account the electron diffusion along the SL axes, tunneling into the BBR, electron capture by localized states and reveres processes of electron detachment. The obtained results are in good agreement with the experimental observations. The employed model determines the capture time, portion of photoelectrons that have been localized by traps in SL, losses of QE and polarization.

2. Experiment

We have studied the pulse response of four photocathodes based on AlInGaAs/AlGaAs SL with strained QWs. All samples were grown on a p-type (100) GaAs substrate by molecular beam epitaxy (MBE). The cathode structure contains a thick $Al_xGa_{1-x}As$ ($x = 0.35 - 0.4$) buffer layer that is p-doped by Be to the level 6×10^{18} cm^{-3}. On the top of the buffer the cathodes working layer was grown containing 4 to 15 periods of $Al_xIn_yGa_{1-x-y}As(a)/Al_zGa_{1-z}As(b)$ SL p-doped to a lower level of 3×10^{17} cm^{-3}. Layer composition, x, y, z, values of the QW- (a) and barrier- (b) layer thickness as well as the number of SL periods (N) are shown in table 1. Above the SL a 6 nm GaAs heavily Be-doped (to the level 7×10^{18} cm^{-3} (1×10^{19} cm^{-3} for SL 5-998)) surface layer was grown to produce thin BBR. Finally the GaAs surface was activated by repeating deposition of cesium and oxygen to achieve the negative electron affinity. All experiments were performed at room temperature.

Table 1. Composition of photocathode working layer: number of SL's periods (N), concentration of Al (x) and In (y) in $Al_xIn_yGa_{1-x-y}As$ QW layer, Al concentration (z) in $Al_zGa_{1-z}As$ barrier layers and their thicknesses (a, b).

Samples	QW		Barrier	Thickness		Periods
	x, %	y, %	z, %	a, nm	b, nm	N
SL 5-998	20	16	28	3.5	4.0	15
SL 6-905	20	15.5	36	5.1	2.3	10
SL 6-908	20	15.5	36	5.1	2.3	6
SL 7-396	20	19	40	5.4	2.1	12

The experimental setup is shown in figure 1. The electron pulses are generated by an approximately 150 fs long laser pulse from a titanium-sapphire laser. An increase of pulse length to 300 fs takes place during optical beam transport to the cathode. These lightpulses are synchronized to the output of a klystron which drives a 2.45 GHz-deflection cavity. By passing the first deflection cavity (quick magnet in figure 1), the longitudinal profile of the electron bunches is transferred into a corresponding transverse profile. The pulse profile can then be measured by moving the electron pulse in the second deflection cavity (slow magnet in figure 1) over a narrow slit and detecting the transmitted current. By analyzing the spin polarization of the transmitted electrons with a Mott polarimeter a time resolved polarization measurement is obtained. Details of the apparatus are described by K. Aulenbacher et al [9].

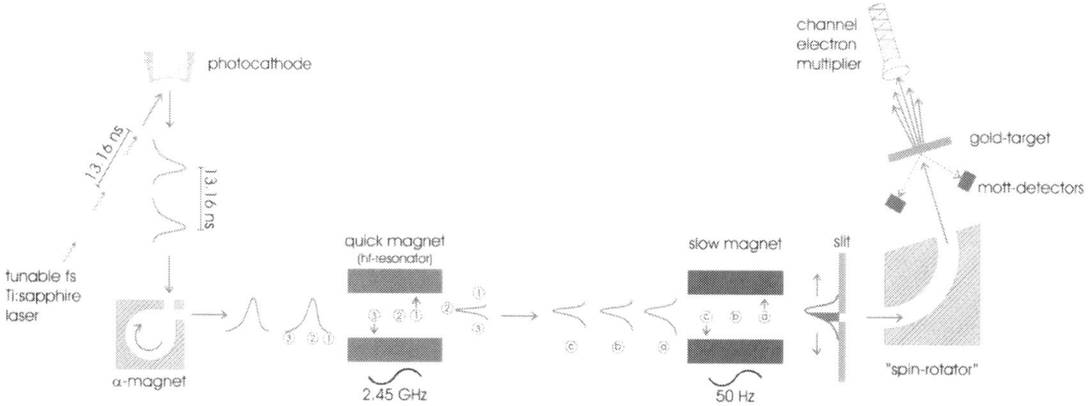

Figure 1. Experimental setup.

3. Theoretical approach

To reproduce the experimentally observed photocathode pulse response we calculate the electron current $I(t)$ from SL to BBR. We neglect the time of electron emission from BBR into the vacuum because this time is smaller than the apparatus resolution time. This fact is proved by pulse response measurements for the cathodes with extremely thin working layers. The electron current $I(t)$ is found by the solution of the time dependent kinetic equation

$$\frac{\partial \hat{\rho}}{\partial t} = -\frac{i}{\hbar}\left[\hat{H}\hat{\rho}\right] + St\{\hat{\rho}\}, \tag{1}$$

where ρ is the electron density matrix, H is the effective electron Hamiltonian for the first electronic miniband $e1$ which describes the quantum electron motion along SL axis. We calculate the miniband energy spectrum using the multiband Kane model, including the conduction band Γ_6, the states of light and heavy holes of the valence band Γ_8 and also the states of the spin-orbit splitted Γ_7 band [12]. The width ΔE of the $e1$ miniband along the SL axis in the considered samples is in the range 32 -37 meV. These values are much smaller than the conduction band offsets which appeared to be larger than 200 meV for all considered samples. Consequently, the vertical electron motion along SL axes can be well described within the tight binding approximation. For effective Hamiltonian H it means that we only have to take into account the coupling matrix element between the neighboring QWs $V \equiv H_{n,n+1} = \Delta E/4$. This matrix element determines the tunneling time between neighboring QWs $\tau_{QW} = \pi\hbar/2V = 2\pi\hbar/\Delta E$ which is about 100 fs for our samples. The total time of free electron motion within SL containing N QWs is N times lager $\tau_{SL} = 2\pi\hbar N/\Delta E$.

The last term on the right hand side of equation (1) is the collision term which takes into account all processes of electron scattering on impurities, phonons etc, photoelectron generation under the optical pumping and electron extraction into BBR. The first contribution to the collision term we write in the constant relaxation time approximation

$$St\{\hat{\rho}\} = -\hat{\rho}/\tau_p. \tag{2}$$

Due to the rather weak coupling between the QWs the momentum relaxation time τ_p is connected mainly with electron scattering in SL's heteroplane within each QW. For the numerical calculation we take $\tau_p = 75$ fs. Note that τ_p is about the time τ_{QW} of electron tunneling between QWs. Thus the vertical electron transport in considered samples has the character of electron diffusion rather than ballistic motion.

The second contribution to the collision term comes from photoelectron generation. Calculation of the photogeneration rate is described in details in Ref. [12]. The third contribution to the collision term is connected with electron tunneling from the last QW to BBR. The corresponding electron current can be written via the number of electrons in the last QW, ρ_{NN}, and tunneling time trough the last barrier, τ_f:

$$I = \frac{\rho_{NN}}{\tau_f}. \tag{3}$$

To calculate τ_f we solve the separate quantum mechanical problem of free electron motion through the single QW to BBR, find the electron density and current and determine eventually τ_f via equation (3). It is worth to note that the obtained τ_f is larger than τ_{QW}, e.g. for SL 5-998 $\tau_f = 0.25$ ps. It is quite natural because the tunneling between the neighboring QWs is a resonant process and its probability is proportional to the first order of the tunneling exponent exp($-\kappa b$), where κ is the electronic wave vector under the barrier and b is the barrier width. The tunneling through the last barrier is a non resonant process and its probability is smaller since it is proportional to the second order of the tunneling exponent exp($-2\kappa b$). Thus the total tunneling time trough the whole SL to BBR is determined by the slowest process, i.e. by the tunneling throw the last barrier.

Equation (3) can be considered as the boundary conditions for the kinetic equation (1) at the heterojunction of SL and BBR. At the opposite SL's side, at the heterojunction with the buffer layer, electron current is zero. In the bulk case the boundary condition is written via the so named surface recombination velocity, $j = n/S$, where n is the electron density. Comparing this with equation (3) we conclude that the surface recombination velocity in our case is equal to

$$S = \frac{d}{\tau_f}, \tag{4}$$

where $d = a + b$ is the SL period. Surface recombination velocities for our SL based cathodes calculated according to equation (4) are smaller than in the case of conventional photoemitters, e.g. for SL 5-998 $S = 3\cdot10^6$ cm/s, for SL 6-905 $S = 5\cdot10^6$ cm/s while in the bulk GaAs $S = 10^7$ cm/s [9].

For numerical simulation of the cathode response we solve the time-dependent kinetic equation (1) with time dependence of the laser pulse intensity described by the Gaussian profile. The obtained electron current $I(t)$ (3) is compared then with experimental pulse response (see next section). The calculated pulse response exponentially decays and we define a corresponding decay time as electron transport time τ_t. It is possible to derive the approximate explicit expression for τ_t. In the steady state under the stationary pumping kinetic equation (1) has an analytical solution if the miniband width ΔE is smaller than the photoelectron energy distribution. The transport time defined in the stationary case as the ratio of number of electrons inside the SL to the generation rate is equal to

$$\tau_t = \frac{\hbar^2 (N-1/2)(N-1)}{6V^2\tau_p} + N\tau_f. \tag{5}$$

According to equation (5), the transport time is a sum of the diffusion time within the SL given by the first term on the r.h.s. of equation (5) and the time $N\tau_f$, taken to penetrate through the last barrier to the BBR. For 12 periods SL 5-998 equation (5) gives the transport time $\tau_t = 5$ ps. The main contribution, 3ps, comes from the tunneling through the last barrier to BBR.

Note that in the limit of thick SL, N >> 1, equation (5) transfers to a corresponding expression of the standard diffusion model

$$\tau_t = \frac{L^2}{3D} + \frac{L}{S}, \tag{6}$$

where $L = Nd$ is the SL's length. The surface recombination velocity S is given by equation (4) and diffusion coefficient D of SL is equal to

$$D = \frac{2V^2 d^2 \tau_p}{\hbar^2}. \tag{7}$$

The typical value of D is about 15 cm^2/s, e.g. for SL 5-998 $D = 12$ cm^2/s and for SL 6-905 $D = 17$ cm^2/s. It is a few times smaller than the diffusion coefficient $D = 40$ cm^2/s in the bulk GaAs.

4. Results and Discussion

The experimentally observed pulse response of SL 5-998 sample is shown in figure 2 together with the results of our numerical calculations. The figure shows the significant discrepancy between experiment and theory. Namely, the calculated emission current exponentially decays with a transport time of τ_t = 6 ps while the experimental signal reveals the non-exponential decay. During the first six picoseconds after excitation the experimental signal decays faster than theoretically predicted with decay time τ_1 = 4 ps. Then its decay slows down and can be fitted by an exponent with a decay time of τ_2 = 13 ps which is larger than τ_t. This picture is similar for all four samples. We can distinguish in the

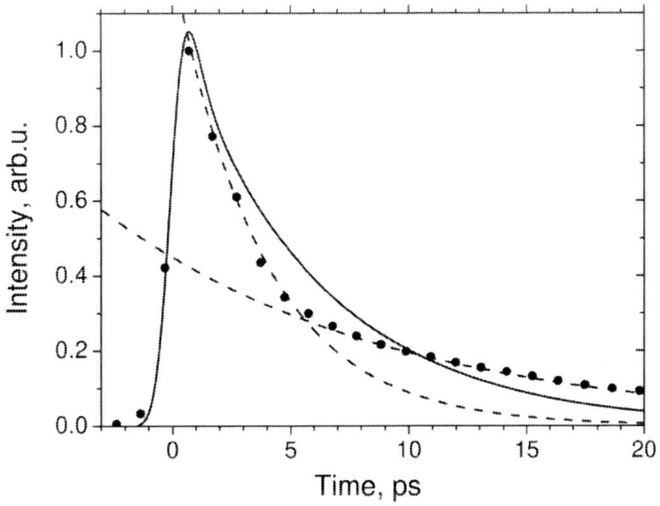

Figure 2. Pulse response of SL 5-998 sample, experimental data is shown by dots and theoretical calculations by solid line. Dashed lines show the exponential fits of experimental data.

cathodes pulse response a fast decay followed by a rather long tail with decay times related to the calculated transport time as $\tau_1 < \tau_t < \tau_2$.

In figure 3 the pulse response of SL 7-396 sample is shown together with polarization of electron emission. The polarization decay is fitted by the exponent with spin relaxation time τ_s= 81ps. Note that electron polarization even 30 ps after the pulse excitation is larger than 60%. It means that the long tail of pulse response is connected with electrons generated in the SL.

These results can be explained by partial electron localization within the SL. Indeed, in the presence of electronic traps the emission current will decrease faster just after the photoexcitation because of combination of two processes depopulating $e1$ miniband states: electron tunneling to BBR and capture by localized states in SL. The existence of long tails in pulse response can also be well understood within the model of partial localization. This slow part of emission current is connected with trapped electrons that return to miniband states due to the thermoactivation detachment processes.

To verify the proposed model we apply the very simple kinetic scheme which takes into account two groups of electron states: delocalized electron states in the $e1$miniband and localized electronic states below the mobility edge. We will describe the rate of the detachment processes by a single detachment time τ_d instead of the broad distribution of τ_d for various localized states. The rate of the electron capture we denote as $1/\tau_c$ and the extraction rate of miniband electrons to BBR is $1/\tau_t$. Then the kinetics of the miniband (delocalized) and localized electrons is described by the following equations:

$$\frac{dn_1}{dt} = -\frac{n_1}{\tau_t} - \frac{n_1}{\tau_c} + \frac{n_2}{\tau_d},$$

$$\frac{dn_2}{dt} = \frac{n_1}{\tau_c} - \frac{n_2}{\tau_d}, \qquad (8)$$

where n_1 and n_2 denotes the number of miniband and localized electrons respectively. To reproduce the decay of pulse response we use the following initial conditions: $n_1 = 1$, $n_2 = 0$, which assumes that on the instant of excitation there are no localized electrons. In this case the solution of equations (8) is given by the sum of two exponents

$$n_1(t) = \frac{(\tau_2 - \tau_t)}{(\tau_2 - \tau_1)}\exp(-\frac{t}{\tau_1}) + \frac{(\tau_t - \tau_1)}{(\tau_2 - \tau_1)}\exp(-\frac{t}{\tau_2}),$$

$$\tau_{1,2}^{-1} = \frac{1}{2}\left(\tau_t^{-1} + \tau_c^{-1} + \tau_d^{-1} \pm \quad \tau_t^{-2} + \tau_c^{-2} + \tau_d^{-2} + 2\tau_t^{-1}\tau_c^{-1} + 2\tau_d^{-1}\tau_c^{-1} - 2\tau_t^{-1}\tau_d^{-1}\right). \qquad (9)$$

The capture and detachment times $\tau_{c,d}$ in equations (8,9) are used as fitting parameters while the miniband transport time τ_t is calculated according to equation (5).

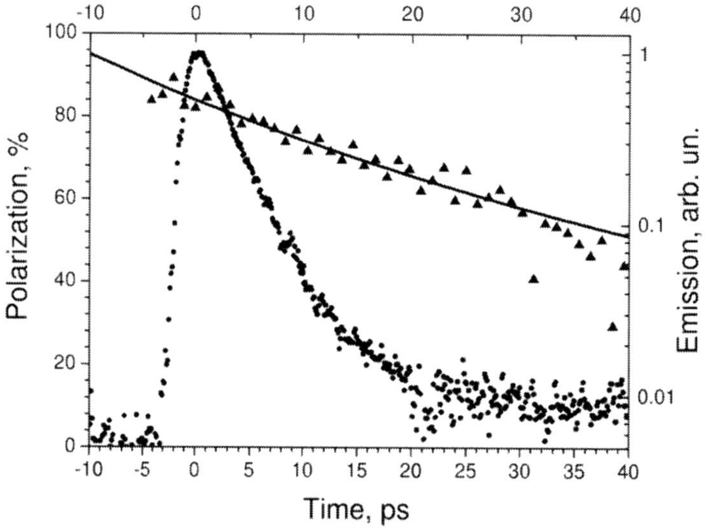

Figure 3. Pulse response of SL 7-396 sample, emission current is shown by dots, polarization of emitted electrons is shown by triangles; solid line shows the exponential fit for depolarization curve.

The results of the proposed model for four samples SL 5-998, SL 7-396, SL 6-905 and SL 6-908 are shown in figure 4 and table 2. The calculated emission current is shown in figure 4 by solid lines and is proportional to the number of miniband electrons $n_1(t)$ given by equation (9). Good agreement between calculated results and experimental data verifies the applicability of the proposed model. Parameters of the model used in the calculations are presented in table 2. The capture and detachment times are chosen in order to match decay times $\tau_{1,2}$ in equation (9) with decay times of experimental signal. The portion of localized electrons which is equal to $\tau_t/(\tau_t + \tau_c)$ shows what part of photoelectrons have been trapped in SL. The average transport time is the total time that electrons spend in SL prior to their extraction to BBR averaged according to equation (9). Finally we calculate the value of electron and polarization losses in SL using the electron life-time τ_r = 50 ps [13] and spin relaxation time τ_s = 81 ps.

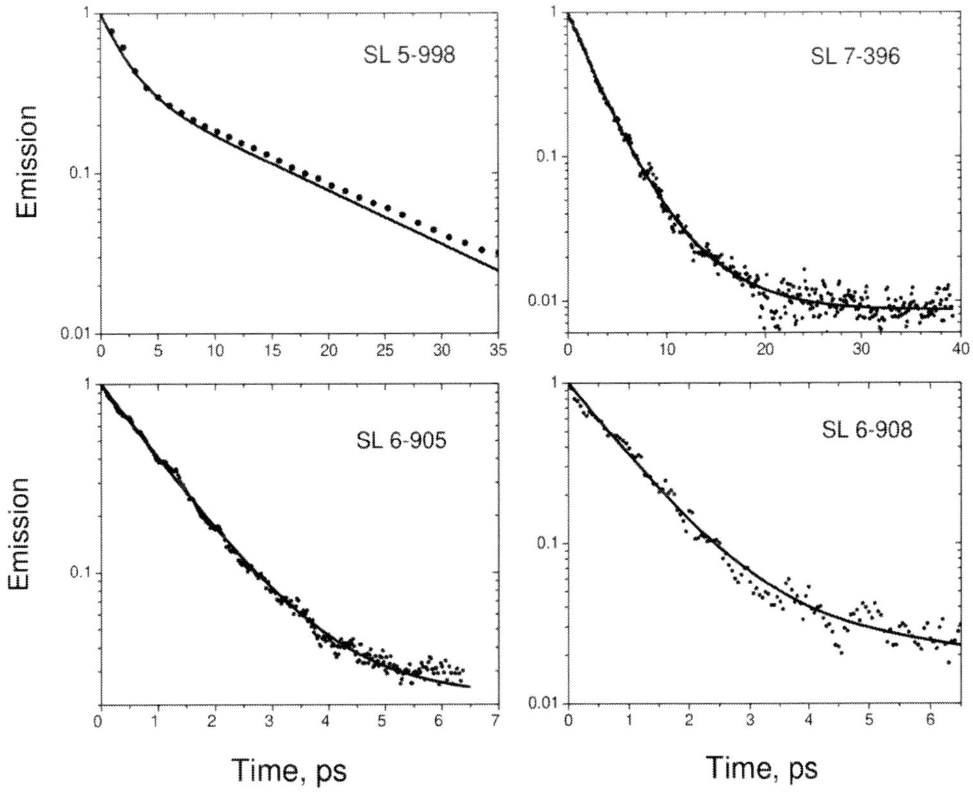

Figure 4. Decay of the pulse response calculated according to equations (9) for SL 5-998, SL 7-396, SL 6-905 and SL 6-908 samples.

Electron localization slows down the electron transport which leads to disappointing losses of QE and polarization. Let us estimate the maximal QE at polarization maximum that could be achieved in the case of ideal SL, i.e. SL without electronic traps. The diffusion length $L_D = (D\tau_r)^{½}$ for SL 6-908 is equal to 0.29 μm which is equivalent $N_D = L_D/d$ = 39 QWs. The maximal QE of the photocathode with BBR as in SL 6-908 sample and with infinite SL with composition identical to SL 6-908 is equal to 2.2%. Photocathode SL 6-908 has the best transport properties of all considered samples. It has the lowest level of electron localization and hence the minimal electronic losses. If it would be possible to grow cathode structure with the same parameters as SL 6-908 and containing $N = N_D$ SL periods its QE will be 1.9%, i.e. three times larger than QE = 0.57% of SL 6-908 sample.

However our study demonstrates that the longer SL has worse transport properties than the shorter one. Indeed, SL 6-905 sample containing ten periods SL with the same composition as SL 6-908 has electronic losses which are six times larger than in SL 6-908 samples. It means that the density of electronic traps and hence the level of electron localization in this sample is much higher than in SL 6-908. Larger density of electronic traps leads to shorter capture time and higher localization level (see table 2). Another important parameter depending on trap density is the ratio of capture and detachment times which is equal to the ratio of miniband and localized electrons at the equilibrium. For SL 6-905 this parameter is smaller by the order of magnitude than for SL 6-908. The large difference of electron localization and corresponding electronic losses in SL 6-905 and SL 6-908 samples explains why the cathodes with ten and six period SLs have close values of QE.

Table 2. Parameters of vertical electron transport, transport time of $e1$ miniband (τ_t), capture time (τ_c), detachment time (τ_d), portion of localized electrons, total transport time, losses of photoelectrons and spin polarization in SL.

Samples	Number of periods	Miniband transport time, ps	Capture time, ps	Detachment time, ps	Partial localiza-tion, %	Average transport time, ps	Electron losses, %	Spin losses, %
SL 5-998	15	6.0	5.6	4.3	52	11	17	9
SL 7-396	12	4.5	9.4	66	32	36	19	11
SL 6-905	10	2.5	1.7	27	60	45	37	11
SL 6-908	6	1.3	4.8	7.6	21	3.4	6	3

Within the present work we do not determine the nature of the localized electronic states as well as possible dependence of their density on SL's thickness. The photocathode structure is heavily p-doped in order to achieve thin BBR. The highest doping level about 10^{19} cm^{-3} is applied in BBR, but the working layer is also heavily doped up to 3×10^{17} cm^{-3}. Fluctuations of impurity potential might be responsible for the formation of localized electron states. Such phenomena have been observed earlier in GaAs/AlAs SLs at a comparable doping level of 1×10^{17} - 7×10^{17} cm^{-3} [11]. The 60% of localized states in the lowest electron miniband $e1$ at a doping level of 4×10^{17} cm^{-3} have been reported. The drop of vertical electron conductivity caused by partial electron localization was observed.

Unavoidable fluctuations of the layer composition and thicknesses can also contribute to partial electron localization. However one can assume that the density of electronic traps formed due to these mechanisms should not depend on the SL's length. On the contrary, the density of structural defects such as dislocations formed in SL due to the strain relaxation increases for thick SLs. Lattice mismatch deformation of the QW layers shifts the conduction band edge upwards by approximately 0.1 eV. Thus the deformation potential of dislocation plays a role of deep well for miniband electrons and can form the localized electron states. Note that the strong deformation of the SL is needed to produce sufficient energy splitting between light- and heavy-hole minibands to achieve high electron polarization. Further work will focus on the nature of electron localization. Its main goal will be to determine the maximal number of SL's periods that can be grown keeping good transport properties in combination with high electron polarization.

5. Conclusions

The spin polarized electron transport in photocathodes based on AlInGaAs/AlGaAs SLs with strained QWs has been studied. The pulse response of the cathodes has been measured and response and spin relaxation times have been determined. We develop the theoretical description of vertical electron transport in SL and compare the obtained experimental data on pulse response with results of our calculations. The analysis performed argues the presence of partial electron localization in SL. The proposed kinetic model taking into account transport of miniband electrons from SL to BBR, their capture by electronic traps and the reverse detachment process provides a good agreement between experimental data and theoretical results. We demonstrate that electron localization slows down electron transport and leads to losses of photoelectrons and polarization. Partial electron localization limits maximal QE and useful thickness of SL based cathodes working layer.

6. Aknowlegments
This work was supported by Russian Ministry of Education and Science under grant 2.1.1/2240 and by DFG through SFB 443.

References

[1] Subashiev A V, Mamaev Yu A, Yashin Yu P, Clendenin J E 1999 *Phys. Low-Dim. Struct.* **1-2** 1
[2] Nishitani T et al 2005 *J. Appl. Phys.* **97** 094907
[3] Mamaev Yu A, Gerchikov L G, Yashin Yu P, Vasiliev D A, Kuzmichev V V, Ustinov V M, Zhukov A E and Mikhrin V S 2008 *Appl. Phys. Lett.* **93**, 81114
[4] Gerchikov L G, Mamaev Yu A, Subashiev A V, Yashin Yu P, Vasil'ev D A, Kuz'michev V V, Zhukov A E, Semenova E S, Vasil'ev A P and Ustinov V M 2006 *Semiconductors* **40**, 1326
[5] Brachmann A, Clendenin J E; Garwin E L, Ioakeimidi K, Kirby R E, Maruyama T, Prescott C Y, Sheppard J, Turner J and Zhou F 2007 *17th International Spin Physics Symposium: AIP Conference Proceedings* (*Kyoto, Japan, 2-7 October 2006*) **915**, pp 1091-4
[6] Groebli J C, Oberli D, Meier F, Dommann A, Mamaev Yu A, Subashiev A V and Yashin Yu P 1995 *Phys. Rev. Lett.* **74** 2106.
[7] Bakarov A K, Jaroshevich A S, Toropov A I, Scheibler H E, Terekhov A S 1999 *Int. Workshop on Polarized Sources and Targets* (*PST 99*): *Proc.* (*Erlangen, Germany, 29 September - 2 October 1999*) p 238
[8] Gerchikov L G, Mamaev Yu A, Yashin Yu P, Vasil'ev D A, Kuz'michev V V, Ustinov V M, Zhukov A E, Vasil'ev A P and Mikhrin V S 2009 *Semiconductors* **43** 463
[9] Aulenbacher K, Schuler J, Harrach D, Reichert E, Roethgen J, Subashev A, Tioukine V and Yashin Ya 2002 *J. Appl. Phys.* **92** 7536
[10] Oskotskij B D, Subashiev A V and Mamaev Yu A 1997 *Phys. Low-Dim. Struct.* **1-2** 77
[11] Pusep Yu A, Chiquito A J, Mergulhao S and Galzerani J C 1997 *Phys. Rev.* B **56** 0163
[12] Subashiev A V, Gerchikov L G, Ipatov A I 2004 *J. Appl. Phys.* **96** 1511
[13] Matsuyama T, Takikita H, Horinaka H, Wada K, Nakanishi T, Okumi S, Nishitani T, Saka T and Kato T 2004 *Jap. J. Appl. Phys.* **43** 3371

A study of an electron affinity of cesium telluride thin film

H Sugiyama[1], K Ogawa[2], J Azuma[2], K Takahashi[2], M Kamada[2], T Nishitani[3], M Tabuchi[4], T Motoki[5], K Takashima[5], A Era[5] and Y Takeda[6]

[1] Accelerator Laboratory, High Energy Accelerator Research Organization (KEK), 1-1 Oho, Tsukuba, Ibaraki 305-0801 Japan.
[2] Synchrotron Light Application Center, Saga University, 1 Honjo-machi, Saga 840-8502 Japan.
[3] The Institute of Physical and Chemical Research (RIKEN), 2-1 Hirosawa, Wako, Saitama 351-0198 Japan.
[4] Venture Business Laboratory, Nagoya University, Furo-cho, Nagoya, Aichi 464-8601 Japan.
[5] Department of Materials Science and Engineering Nagoya University, Furo-cho, Nagoya, Aichi 464-8603 Japan.
[6] Department of Crystalline Materials Science, Nagoya University, Nagoya, Aichi 464-8603 Japan.

E-mail: harues@post.kek.jp

Abstract. The photoemission spectroscopy using synchrotron radiation has been carried out to study the electron affinity of cesium telluride thin films. We observed a formation of NEA surface on p-type gallium arsenide deposited with cesium telluride. An ultra thin film of cesium telluride is discussed as a possible candidate forming an NEA surface. The electron affinity of cesium telluride thin film was found to increase with time after fabrication. An increase of the affinity resulted in deterioration of a quantum efficiency under laser light exposure with constant wavelength. The surface of photocathode in RF gun can suffer enormous damage easily at the conditioning process of RF gun before electron extraction. A method to refresh the photocathode surface after RF conditioning has been required. A photodetachment process is a candidate to refresh aged/damaged photocathode with X-ray exposure. A recovery of quantum efficiency for the aged photocathode film by this refreshing process is shown.

1. Introduction

A cesium telluride thin film, which has a high quantum efficiency and a long working lifetime, has already been established and used as a photocathode material for RF gun in practical use [1]. A cesium telluride photocathode can provide a longer lifetime under severe conditions, poor vacuum condition and higher electric field, comparing to conditions of a DC gun used as a polarized electron source with an NEA (Negative Electron Affinity) photocathode [2]. Therefore we have studied the electron affinity of cesium telluride in order to obtain hints to improve a lifetime of NEA photocathodes by the photoemission spectroscopy using synchrotron radiation [3]. The conventional NEA method uses oxygen and cesium [4]. Oxygen belongs to the same VIth group in the periodic table same as tellurium. The study about the electron affinity of cesium telluride is expected a helpful way for the research to a durability of NEA surface modification. An electron affinity, which is defined by a difference between a vacuum level and a conduction band minimum, has been found to increase with time after fabrication in cesium telluride thin film. An increase of this affinity results in a deterioration of quantum efficiency under laser light exposure with constant wavelength as shown in Fig. 1. Thus the working lifetime of photocathodes has been associated with an increase electron affinity.

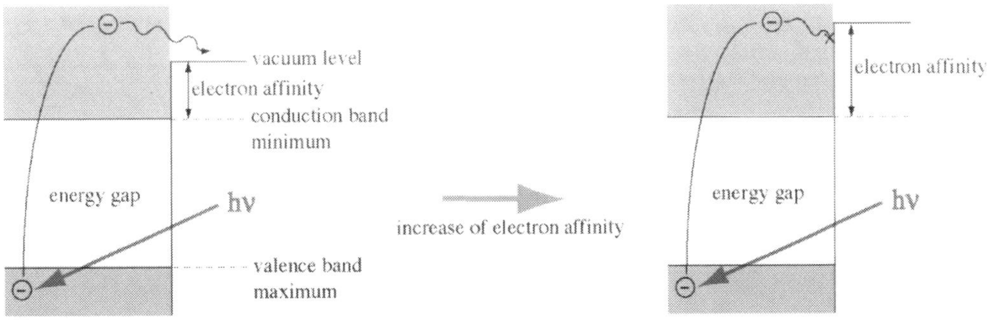

Figure 1. Illustration of a deterioration of quantum efficiency with conceptual energy diagram.

The results of photoemission spectroscopic study using synchrotron radiation were shown in this report. An ultra thin film of cesium telluride is discussed as a possible candidate forming an NEA surface. We observe a construction of NEA surface on p-type gallium arsenide deposited with cesium telluride. And also, photodesorption is discussed as a refresh treatment for a wasted cesium telluride thin film photocathode. The surface of photocathode in RF gun can suffer enormous damage easily at the conditioning process of RF gun before electron extraction. A method to refresh the photocathode surface after RF conditioning is therefore highly desirable. A photodetachment process is proposed to be a candidate to refresh aged/damaged photocathode with X-ray exposure.

2. Experiment

This study has been carried out using the Saga University beamline BL13 at SAGA Light Source (SAGA-LS) [5]. The beamline BL13 consists of two end stations/branches. The one is the VLS station for the high-resolution photoelectron spectroscopy with undulator radiation. Another is the station for the vacuum ultraviolet photoelectron spectroscopy in the energy range from UV to EUV, where the UVSOR-type plane grating monochromator (PGM) has been adapted. At the end station of BL13-PGM, an analyzer of photoelectron spectroscopy which has a resolution of 100 meV at 40 eV photon energy, a system for measurement of total yield, an analyzer of Auger electron spectroscopy, and a LEED system are all set-up in the analyzer chamber which has a base pressure of 2×10^{-8} Pa. Cesium and tellurium sources, an oxygen gas source, a thickness monitor, a sputtering ion gun, a system of sample annealing, and so on are installed in the preparation chamber, which has a base pressure of 2×10^{-8} Pa. The end station has also an air-lock chamber that makes it possible to install samples without a breaking UHV of analyzer and preparation chamber.

3. NEA surface using Cs/Te deposition

The result of cesium (Cs) and tellurium (Te) deposition on p-type gallium arsenide (GaAs) is shown in this section. The p-type GaAs sample fabricated with As cap at Venture Business Laboratory of Nagoya University had a contamination free surface originally. The sample was used for another experiment for NEA surface modification, and after finishing the experiment, the sample was exposure in the air. Hence the sample contaminated with the air was used in this experiment. The Auger electron spectrum of the sample, p-type GaAs deposited with Cs/Te, is shown in figure 2. The peaks from Ga, As, Cs, Te, oxygen (O) and carbon (C) are observed in the spectrum. O and C peaks come from the surface p-type GaAs, which are also observed before deposition Cs and Te. The thickness of Cs/Te deposition is very thin as materials of underlying substrate are observable.

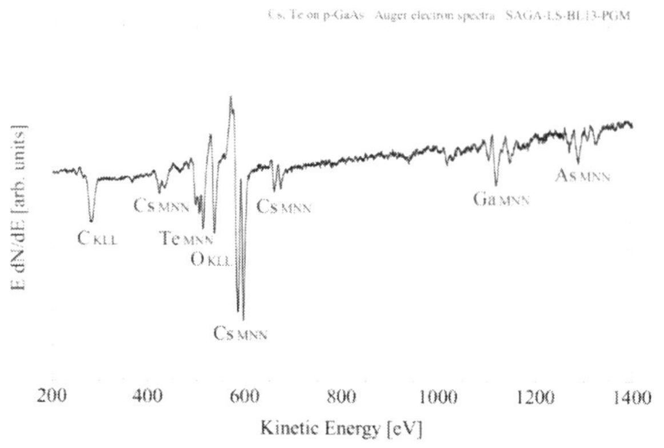

Figure 2. Auger electron spectrum of p-type GaAs deposited with Cs/Te.

The photoemission spectrum around a low energy cut-off is formed from energy distributions of secondary electrons and photoelectrons. The cut-off energy of the spectrum corresponds to the vacuum level of the sample. The photoemission spectrum around a low energy cut-off and around valence band spectrum of the sample before/after Cs and Te deposition is shown in figures 3-4 and 5-6 respectively. The measurement of the low energy cut-off was carried out under negative bias at - 25 Volts in order to collect emitted electrons into the analyzer effectively. In Fig 3, the point at 25 eV corresponds to the work function of the analyzer. The sample work function, which is defined a difference between a vacuum level and the Fermi level, indicates the declination for - 0.2 eV to the work function of the analyzer. In Fig. 4, the work function of the sample decreases for 3.1 eV from that before Cs and Te deposition.

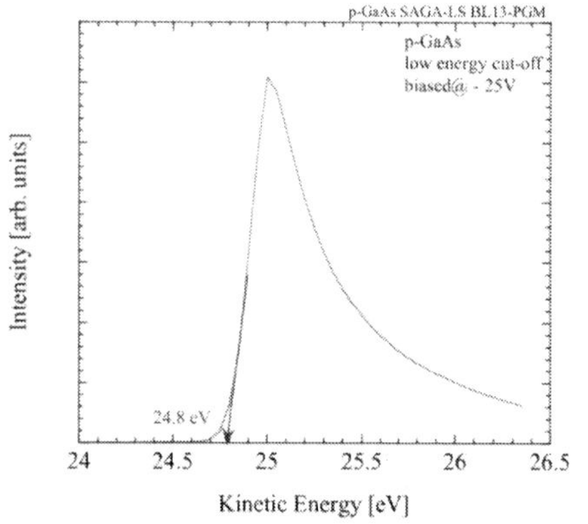

Figure 3. The photoemission spectrum around a low energy cut-off before Cs and Te deposition.

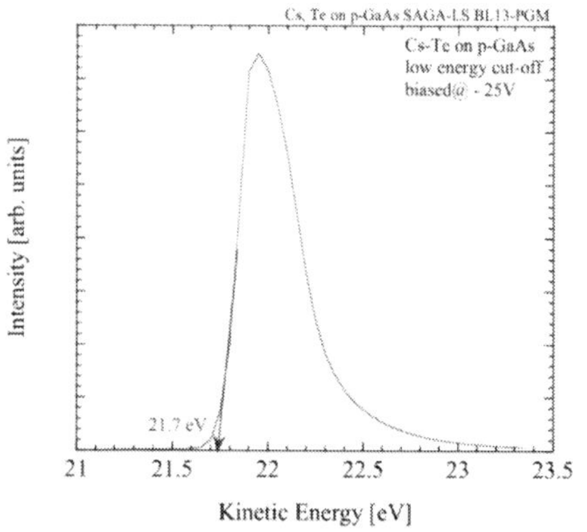

Figure 4. The photoemission spectrum around a low energy cut-off after Cs and Te deposition.

In Figs. 5-6, E_F indicates the Fermi level measured from evaporated gold sample. This p-type sample was highly doped and electrons are filled in the states close to the Fermi level as shown in Figs. 5-6. In Fig. 5, the peak from Ga-3d is observed clear, and the broad peak around 7 eV of binding energy indicates oxidation. In Fig. 6, the peak from Ga-3d still is observed after Cs and Te deposition, other peaks appear as well as a cesium telluride thick film. Since no change the valence band maximum was observed by the Cs/Te deposition, no change to the conduction band minimum might be considered. Therefore the observed shift of the cut-off energy due to the Cs/Te deposition can be recognized as the change of the electron affinity.

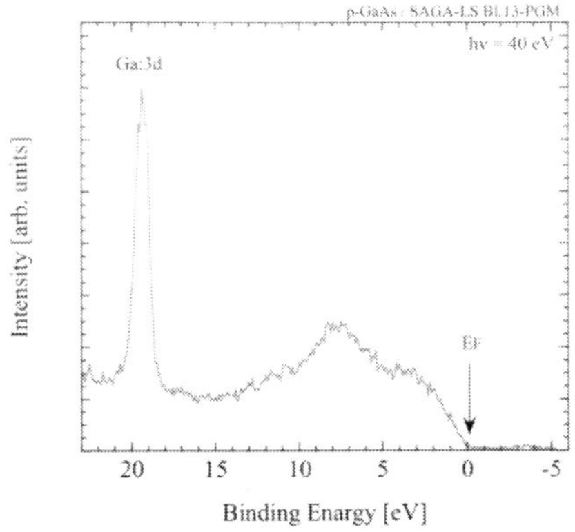

Figure 5. The valence band spectrum before Cs and Te deposition.

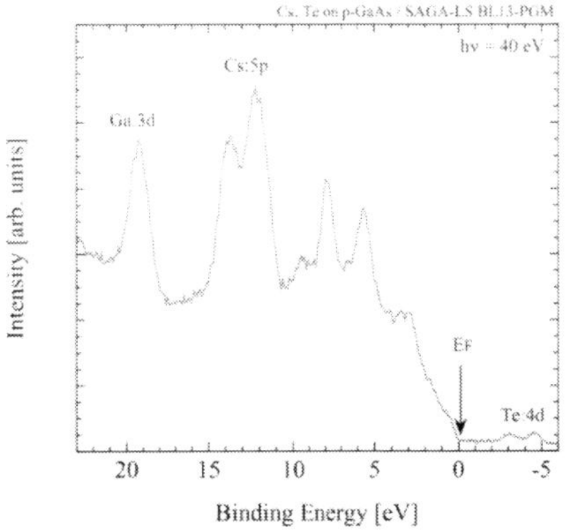

Figure 6. The valence band spectrum after Cs and Te deposition.

An energy diagram summarizing results from the photoemission spectroscopy is shown in Fig. 7. All values shown in the Fig. 7 is measured value in this experiment except an energy gap of 1.4 eV to a GaAs, which is referred from the article [6]. The work function of the sample, can be derived from the work function of the analyzer at 4.6 eV, shifted down for 3.1 eV due to the Cs/Te deposition, and became to 1.3 eV from 4.4 eV. An energy gap of a GaAs semiconductor is about 1.4 eV. Consequently the sample shows a negative electron affinity of - 0.1 eV from this measurement.

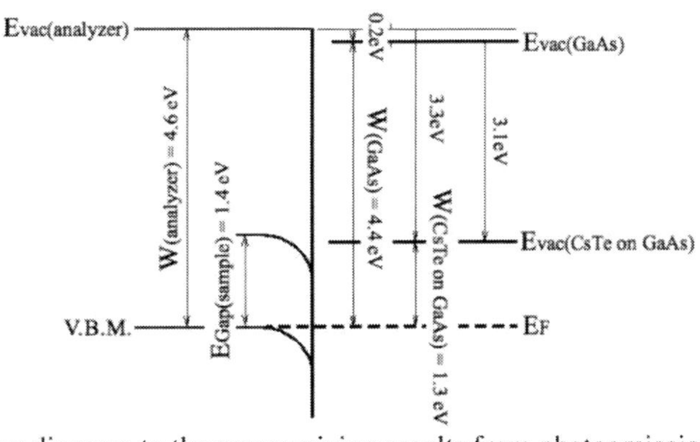

Figure 7. Energy diagram to the summarizing results from photoemission spectroscopy.

4. Photodesorption effect

Figs. 8-9 shows the photoemission spectra of the sample of cesium telluride film using molybdenum as a substrate around a low energy cut-off were measured in biased to – 25 V. As shows in Fig. 10, the electron affinity becomes higher as a function of time from the fabrication. In Figs. 8-9, the red line shows the spectrum for 39 minutes from fabrication, the blue line for 2 hours 15 minutes, the black line for 40 hours 30 minutes, the green line for 138 hours, the sky blue line for 187 hours 30 minutes, the violet line for 188 hours and the grey line for 190 hours respectively. The sample used in this measurement was stored in the air-lock chamber from 150 hours to 187 hours. The vacuum of the air-lock chamber is kept by only turbo-molecular pump. Water molecule attachment under limited vacuum condition might result in a significantly high electron affinity at 187 hour 30 minutes. The spectrum was re-measured after 15 minutes irradiation of unselected light from synchrotron (0th light) at the surface of sample. Then a recovery of the affinity by 0.2eV (the affinity becomes 22.5 eV after irradiation from 22.7 eV before) is observed. This recovery might be explained by a photodesorption effect that detached water from the film by obtaining energy from radiated photons. The further irradiation of 90 minutes gives a increase of the affinity, which might be due to the high intensive photoemission introduced pollution on the surface.

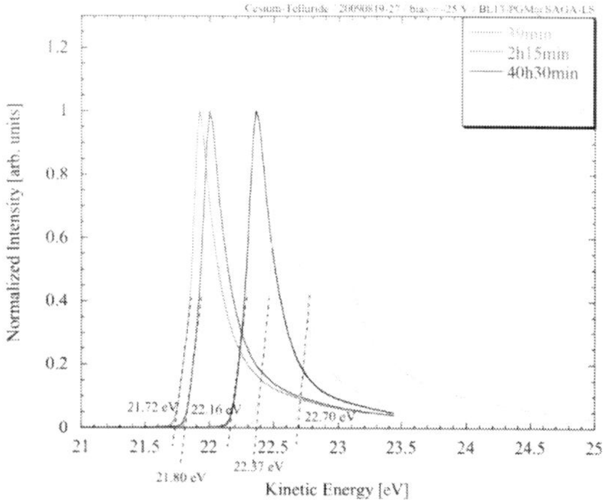

Figure 8. Photoemission spectra of the cesium telluride sample around a low energy cut-off before exposure of 0th light.

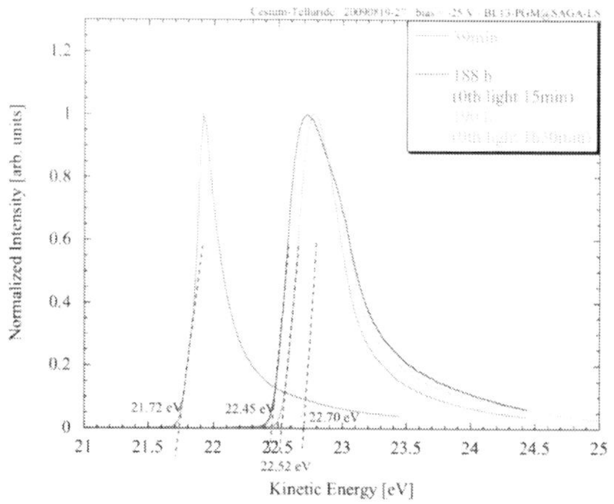

Figure 9. The spectra before/after exposure of 0th light.

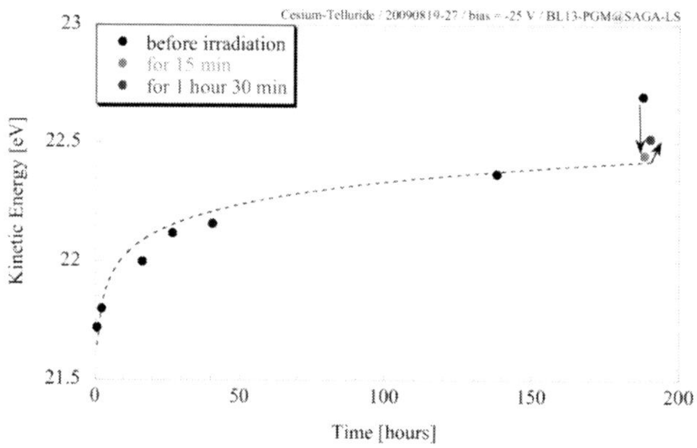

Figure 10. The change of the cut-off energy as a function of time from the sample fabrication. The dotted line is fitted log function to the data as just a guide to eye.

An RF gun needs RF conditioning after an insertion of a cathode plug where cesium telluride is deposited on. This RF conditioning inevitably gives pollution on the cathode surface before gun operation. Refreshment by photodetachment process is a promising method to recover damaged photocathode after RF conditioning. An integration of X-ray tube into RF gun is expected in future study.

5. Summary

The photoemission spectroscopy using synchrotron radiation has been carried out to study the electron affinity of cesium telluride thin films. NEA surface is observed on p-type gallium arsenide deposited with cesium telluride. Photodesorption is a useful method to refresh a wasted surface.

References

[1] for example, Terunuma N, Murata A, Fukuda M, Hirano K, Kamiya Y, Kii T, Kuriki M, Kuroda R, Ohgaki H, Sakaue K, Takano M, Takatomi T, Urakawa J, Washio M, Yamazaki Y, Yang J 2010 *Nucl. Instrum. Methods* A **613** 1

[2] Yamamoto M, Konomi T, Okumi S, Nakagawa Y, Yamamoto N, Tanioku M, Jin X, Ujihara T, Takeda Y, Furuta F, Matsumoto H, Yoshioka M, Kuriki M, Shonaka C and Nakanishi T 2009 *AIP Conference Proceedings* **1149** 987

[3] Sugiyama H, Ogawa K, Azuma J, Takahashi K and Kamada M 2009 *AIP Conference Proceedings* **1149** 1067

[4] Turnbull A. A and Evans G. B 1967 *J. Phys* D **1** 155

[5] Sugiyama H, Takahashi K, Ogawa K, Azuma J, and Kamada M 2010 *AIP Conference Proceedings* **1234** 685

[6] Lautenschlager P, Garriga M, Logothetidis S, and Cardona M 1987 *Phys. Rev.* B **35** 9174

Journal of Physics: Conference Series **298** (2011) 012015 doi:10.1088/1742-6596/298/1/012015

L-band Photocathode RF gun at KEK-STF

H Sugiyama[1], Y Takahashi[2], H Hayano[1], J Urakawa[1], S kashiwagi[3], G Isoyama[4], R Kato[4], N Sugimoto[4] and M Kuriki[5]

[1] Accelerator Laboratory, High Energy Accelerator Research Organization (KEK), 1-1 Oho, Tsukuba, Ibaraki 305-0801 Japan.
[2] The Graduate University for Advanced Studies, School of High Energy Accelerator Science, 1-1 Oho, Tsukuba, Ibaraki 305-0801, Japan.
[3] Research Center for Electron Photon Science, Tohoku University, 1-2-1 Mikamine, Taihaku-ku, Sendai 982-0826 Japan.
[4] Institute of Scientific and Industrial Research (ISIR), Osaka University, Mihogaoka 8-1, Ibaraki, Osaka 567-0047 Japan.
[5] Graduate School of Advanced Sciences of Matter, Hiroshima University, 1-3-1 Kagamiyama, Higashishiroshima, Hiroshima 739-8530 Japan.

E-mail: harues@post.kek.jp

Abstract. The superconducting RF test facility (STF) in KEK is a facility to promote R&D of the International Linear Collider (ILC) cavities and cryomodule. L-band photocathode RF gun has been developed at KEK-STF as an electron beam source for cryomodule test scheduled in autumn of 2011. The RF cavity of the gun will be operated with a 1.3 GHz RF frequency, 1 msec RF pulse width, 5 Hz repetition rate at normal conductivity. The cavity was prepared by collaborative work with DESY and FNAL, and fabricated by FNAL. The RF conditioning of the cavity has been started since April 2010. A cesium telluride thin film as a photocathode material has been adopted, and the preparation equipment for cesium telluride has been newly designed and constructed. By using this new system, a fabrication and a performance estimation of the cesium telluride thin film as a photocathode are the next step of the research.

1. Background

The superconducting RF test facility (STF) in KEK is a facility to promote R&D of the International Linear Collider (ILC) cavities and cryomodule [1]. In STF, the "S1-Global cryomodule experiment" which is aiming to achieve ILC high gradient cavity performance by the international collaborative effort using DESY, FNAL and KEK cavities, is under execution [2]. After the S1-Global experiment finished in 2010, the construction of STF phase 2 accelerator will start in 2011. The phase 2 accelerator consists of an RF gun electron source, superconducting capture cavities, and ILC cryomodule. An L-band photocathode RF gun that will be used there has been developed at KEK-STF as a low emittance and high charge electron source, which has ILC, beam structure for charge and bunch spacing.

2. Overview of RF gun at KEK-STF

The cavity of RF gun has 1.5 cells and a resonance frequency at 1.3 GHz was fabricated by FNAL under KEK-FNAL collaboration, and the used design was based on DESY RF gun under DESY-FNAL collaboration. The RF gun used a photocathode on the cathode plug in the cavity end plate. The cathode plug is inserted from the backside of the cavity end plate. And a cesium telluride thin film is deposited on the cathode plug. The cesium telluride preparation chamber is connected in the back of the cavity. The solenoid magnets are placed around the cavity. To cancel the magnetic field on the point of cathode, there is a backing coil in addition to a main coil in the solenoid. A ceramics RF window is used in the waveguide to keep vacuum in the cavity. There are ion pumps connected to the waveguide: one is 100-litre/sec evacuation and the two are 200 litre/sec evacuations for the pumping of the cavity. Also there is an ion pump of 200 litre/sec evacuation for the preparation chamber. During a deposition of cesium telluride thin film, the cavity can be separated from the preparation chamber by a gate valve. The cathode plug is movable between the cavity end plate and the preparation chamber by the motor driven push rod behind the preparation chamber.

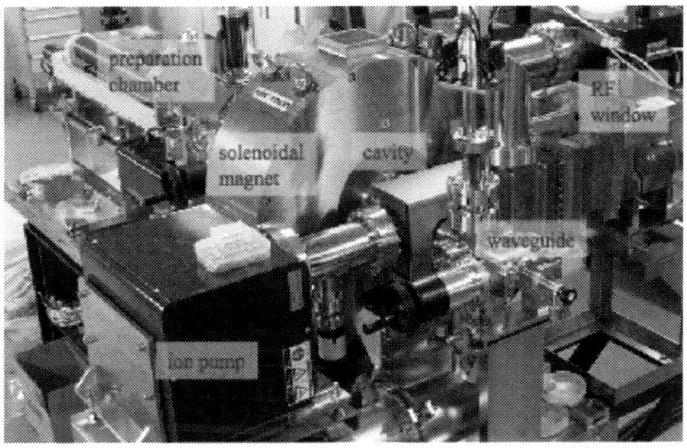

Figure 1. Overview of L-band RF gun at KEK-STF

3. RF conditioning of the gun cavity

An RF conditioning of the cavity was carried out at up to 1.7 MW of PF power, 1 msec of the pulse width, 5 Hz of repetition rate, from April to May of 2010. The conditioning is performed to keep the vacuum in the cavity less than 10^{-6} Pa when the power reached at 1.7 MW. Fig 2 shows a conditioning history of a day for a power of 1.7MW. The green line shows forward RF power into the cavity, and the sky blue line is for the reflected RF power. The reflection was kept well small as the forward power went to 1.7 MW. The red line shows vacuum level of the cavity. It was also kept well below 10^{-6} Pa with spike sometime, however, the vacuum level of the preparation chamber (blue line) was not spiky.

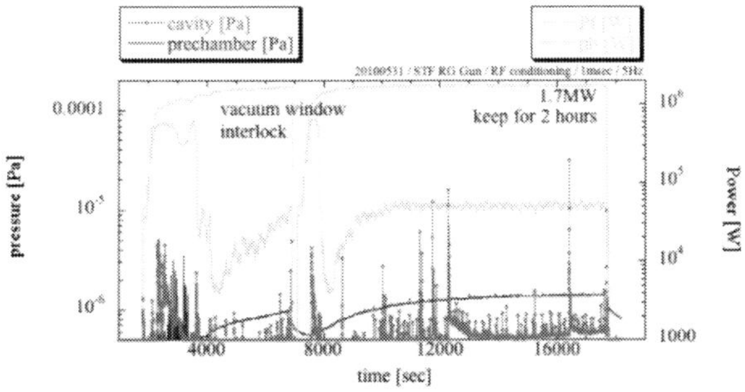

Figure 2. A conditioning history of a day

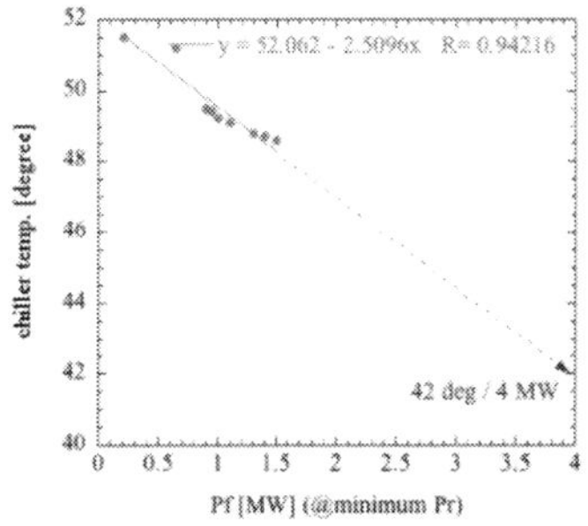

Figure 3. Chiller water temperature vs RF Power.

The RF power was tuned on with 1.3 GHz at 51.5 degree C of the cooling water temperature. However, a detuning of the cavity occurred due to a thermal action by applied RF power. Therefore the set point of the temperature of the chiller was required to let down as a power rise. The targeted power 3.5 MW (corresponding to 40 MV/m on the cathode surface) needs 43 degree C of the chiller water temperature as shown in Fig 3. The chiller that is used seems to be sufficient to reach the target RF power. The conditioning of the cavity in now suspended to carry out the S1-Global experiment. The resuming of the conditioning is scheduled on April of 2011.

4. Cesium telluride preparation chamber

The cesium telluride preparation chamber is connected to the RF cavity backside directly. The motor driven manipulator for a transfer of the cathode plug to the RF cavity has a stroke of 500 mm. The cesium telluride thin film is fabricated on the surface of the cathode plug directly in the preparation chamber. The tellurium and cesium sources are mounted on a vertical manipulation shaft in series. The thickness of a tellurium film can be maintained with monitoring by a quartz thickness meter (INFICON, SQM160) and with a current control of a heating wire for tellurium. A cesium deposition can be carried out in control by a monitoring a photocurrent. The quantum efficiency is measured using an UV-LED (cental wavelength on 261 nm, PicoQuant, PLS 265-10) and a pico-ammeter incorporated a voltage source (> +/- 500 V, Keithley, 6487). The in-situ measurement for the quantum efficiency of the cesium telluride thin film photocathode can be carried out in the preparation chamber after the fabrication, and then the film on the surface of the plug can be installed to the RF cavity without air exposure. The wasted thin film can be removed by a heating wire inside the plug at a temperature of 550 degrees C. The temperature can be monitored using a radiation thermometer (IMPAC, IPE140) that has a range from 30 to 1000 degrees C through a sapphire glass. The commissioning of the preparation chamber is underway for cathode plug cleaning and for cesium telluride deposition.

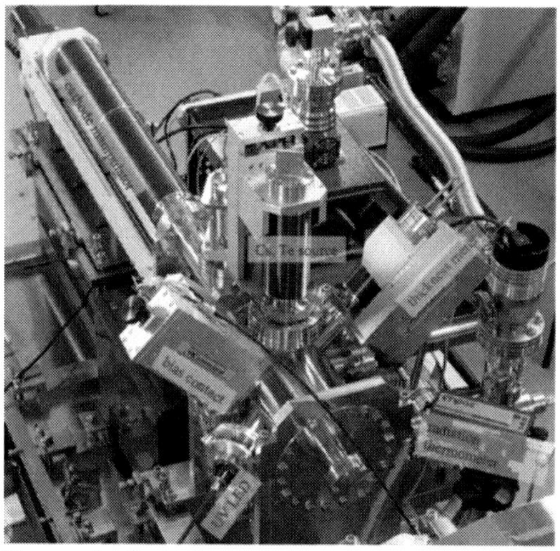

Figure 4. Overview of the preparation chamber.

Figure 5. Inside of the preparation chamber.

5. Cathode plug

The cathode plug and the cathode periphery equipments are required to be made from non-magnetism materials. The cathode plug is made from a molybdenum block. The molybdenum has a large reflectivity for an UV light. A part of the UV light goes through out the thin film, and then reflected light on the molybdenum surface goes back to the thin film again. A molybdenum surface is suitable for the substrate of the thin film photocathode with UV light driven. In this system, the wasted thin film will be cleared off using a heating up the cathode plug at 550 degree C directly. Usually an RF contact of the cathode plug is made from beryllium copper alloy. However, the contact that made from beryllium copper alloy is lost elasticity by heated up to around 400 degrees C. Hence, in this system, a beryllium copper as a material of RF contact cannot be adopted. Considering this effect, we adopt an Inconel750X that is high temperature tolerant for a material of an RF contact.

6. Summary

The L-band photocathode RF gun has been developed at KEK-STF for beam source of the cryomodule test. The 3.5 MW RF power is targeted for beam operation, and now is conditioned up to 1.7 MW. The conditioning will resume on April 2011. The cesium telluride preparation chamber and the cathode plug are already fabricated. The fabrication and the performance estimation for the cesium telluride thin film as a photocathode are under commissioning.

Acknowledgements

This research was partially supported by the Ministry of Education, Science, Sports and Culture, Grant-in-Aid for Scientific Research(C)-226040140017.

References

[1] STF group *"STF Phase-1 Activity Report"* 2009 *KEK Report 2009-3 April*
[2] Ohuchi N, Kako E, Kondou Y, Hayano H, et al., 2010 *Proceedings of IPAC'10* 3356

Development of spin-polarized transmission electron microscope

M Kuwahara [1], **Y Takeda** [2], **K Saitoh** [1,2], **T Ujihara** [2], **H Asano** [2], **T Nakanishi** [3], **N Tanaka** [1,2]

[1] EcoTopia Science Institute, Nagoya University, Nagoya, Aichi 464-8603, Japan

[2] Graduate school of engineering, Nagoya University, Nagoya, Aichi 464-8603, Japan

[3] Graduate school of science, Nagoya University, Nagoya, Aichi 464-8602, Japan

E-mail: kuwahara@esi.nagoya-u.ac.jp

Abstract. In order to study spin related phenomena in nano-size materials, spin-polarized electron source (PES) has been employed for the incident beam in transmission electron microscope (TEM). The PES has been designed and constructed with optimizing for spin-polarized TEM. The illuminating system of TEM is also designed to focus the spin-polarized electron beam emitted from a semiconductor photocathode with a negative electron affinity (NEA) surface. The beam energy is set to below 40 keV which is lower energy type as a TEM, because the spin interaction with condensed matters is very small corresponding with a Coulomb interaction. The polarized electron gun has realized in an extra high vacuum (XHV) condition and high field gradient of 4 MV/m on a surface of photocathode. Furthermore, it demonstrated that 40-keV polarized electron beam was operated with a sub-milli second pulse mode by using the backside excitation type photocathode. This high performance PES will make it possible to observe dynamically a magnetic field images with high contrast and high-speed temporal imaging in TEM.

1. Introduction

The polarized electron beam plays important roles in high-energy physics and materials sciences related to electron spin. The polarized electron source (PES) technology has been developed intensively for accelerators, and recently it was also applied to materials science instrumentations to observe the surface magnetic phenomena [1-4]. Furthermore the electron spin plays very important roles in not only fundamental physics but also some key technologies such as spintronics and quantum-information technology [5-7].

We have started the research and development of PES for applying to a transmission electron microscope (TEM) [8]. The spin-polarized electrons can be generated by using an optical orientation of the proper III-V type semiconductors and a extraction mechanism into vacuum using a negative electron affinity (NEA) surface, as explained by the following three-step model. First, the circularly polarized laser light illuminates the semiconductor and spin-selectively excites the valence band electrons to the conduction band. Secondly, these polarized electrons in the conduction band diffuse and some of them drift to the surface region. Finally, these polarized electrons are extracted into vacuum through the NEA surface under applying a negative electrostatic field. To realize the polarized electron beam experiments in the future International Linear Collider (ILC) project, the PES technology has been developed intensively from a middle of 80's [9]. As results, several beam performances were drastically improved, which include electron spin polarization (ESP), quantum efficiency (QE), a sub-nanosecond multi bunch structured beam and low beam emittance. The high ESP of 92 % and high QE of 0.5 % were already realized by a GaAs-GaAsP strained superlattice (SL) photocathode with the active layer with of 100 nm and the excitation wavelength of 780 nm [10, 11]. It was also demonstrated that the normalized thermal beam emittance can be minimized to as low as 0.15 π mm mrad.

This high performance PES will make it possible to dynamically observe magnetic structure images with high-resolution spatial imaging and high-speed temporal imaging in TEM. Then we have started the development of spin-polarized TEM (SPTEM). In this paper, we would like to share an overview of the PES and the pulse performance. We will also report about the construction of the new PES for TEM.

2. Overview of spin-polarized TEM

The overview of the SPTEM system is shown in figure 1. The SPTEM consists of a PES, illumination-lens system including spin manipulator, objective-lens system, projector-lens system and image-detection system. The polarized electron beam emitted from the PES first passes through a spin-rotator and a condenser-lens system in order to concentrate the beam on the specimen. Then the beam passes through the specimen. Some of the electrons pass all the way through, and others are scattered by some interactions that are the Coulomb interaction, spin-orbit interaction and spin-spin interaction. The modified beam then passes through the objective lens, the projector-lens and onto a fluorescent screen where the final image is observed. Hence the electron beam passes entirely through the specimen, the diffraction pattern gives the observed comprehensive view of the interior of the specimen including magnetic distribution. In the whole parts of vacuum system in the SPTEM excepting for the detection chamber, these vacuum conditions are maintained in ultra-high vacuum (UHV) condition to avoid a destruction of an NEA surface due to the residual gas. Especially, the illumination-lens system was newly developed in this project, which consists of a spin-rotator for transverse rotation (SRB), and a condenser lens (CL) system including in-plane spin-rotation lens magnet (SRL). A 90-degrees bending system was employed for the SRB to apply both an electric field and a magnetic field for rotation in any azimuth angle. The system also can realize to correct the over-rotated angle due to the anomalous magnetic moment of electron and the Thomas precession in the

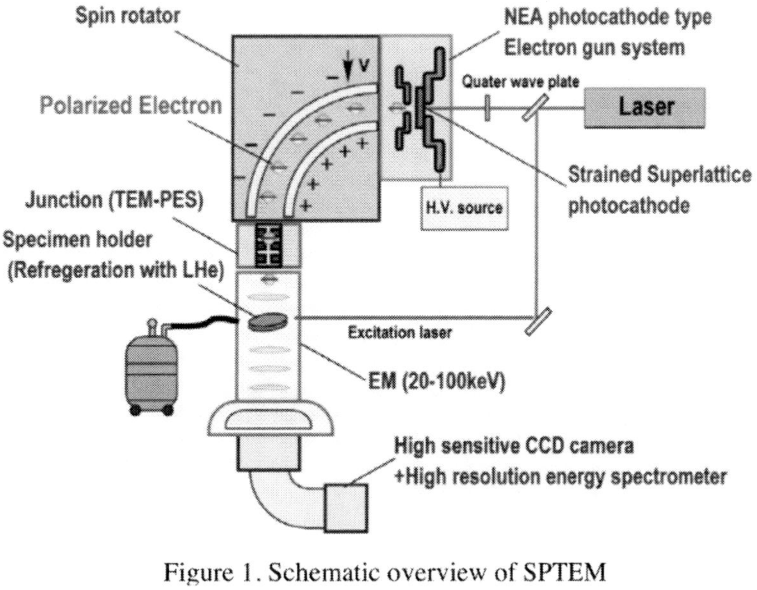

Figure 1. Schematic overview of SPTEM

relativistic effect. The electron beam optics in the PES and the illumination lens system is shown in figure 2.

3. Polarized electron source

3.1. Polarized electron gun

The PES is required to produce high polarization and high brightness beam simultaneously. The 40-keV PES system was newly designed and built to install the transmission type photocathode instead of ordinal LaB_6 electron gun. The PES consisted of three components in vacuum chambers; 40-kV DC-gun, a differentially pumping chamber equipping a collector lens and a photocathode preparation chamber. The photocathode preparation chamber connected to the 40-keV gun is used for activation of photocathode surface to create NEA surface. In order to maintain the NEA surface for long time, two kinds of advanced techniques; the ultrahigh vacuum and the dark current (between the high field gradient gun-electrodes) reduction are absolutely required. The PES apparatus is designed to satisfy the vacuum condition of UHV. Especially the vacuum around the photocathode is kept in XHV (~several$\times 10^{-10}$ Pa) condition to protect the NEA surface from (1) adsorption of the residual gas and (2) positive ion back-bombardment caused by beam collisions with gas molecules. Each chamber are evacuated by ion pumps and non-evaporable getter (NEG) pumps. The components of 40-kV gun are shown in Figure 3. The cathode and anode are made of molybdenum and titanium respectively, since it showed the best performance in our test experiment, where the dark current was suppressed below 1nA under applied field gradient of 134 MV/m on the molybdenum cathode surface [12]. As the final metal surface polishing process, both procedures of electrochemical-etching and fine buffering were simultaneously employed to obtain a mirror-flat surface [13].

An activation process of NEA surface is carried out in a preparation chamber. A photocathode sample is attached to the molybdenum cathode-pack stage, and heated to remove the surface arsenic layer to obtain clean surface by a PBN (pyrolitic boron nitride) heater. Cesium is evaporated from a commercially available dispenser which is placed in front of the photocathode by using a linear manipulator during the cesiation process. Oxygen is introduced into the preparation system from a pure oxygen bottle through a variable leak valve. Circular light is led to the gun by an optical fiber and finally focused on to the photocathode surface from the backside through the focusing lens. A high field-gradient for acceleration is necessary to suppress the space-charge-dispersion effect to keep the high beam brightness. This gun could supply the high gradient of 4 MV/m to the photocathode surface, since a gap separation of electrodes and a supplied voltage are 8.0 mm and 40-kV, respectively. In this condition the dark current in the vacuum still does not exceed a level of 0.1nA. The PES, after the 40-

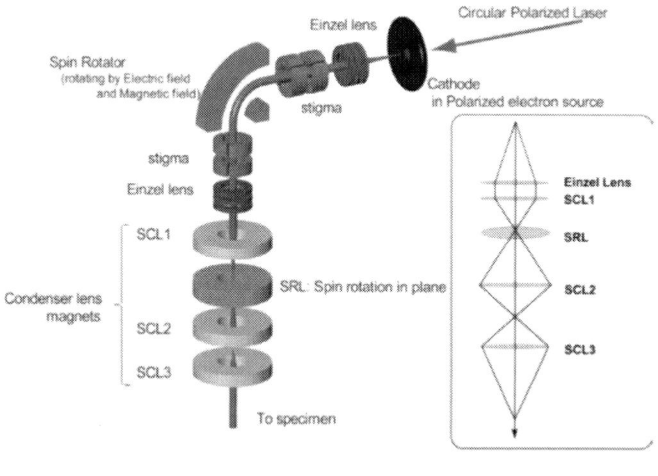

Figure 2. Electron beam optics in the spin-rotator and the condenser lens system of SPTEM

keV gun constructed and tested, was connected to a beam diagnosis system. The figure 4 shows the temporal current profile of polarized electron beam emitted from the 40-keV PES and a phosphorous screen image of the spatial profile using a transmission-type photocathode. It was demonstrated that the PES could emit polarized electron beam using a transmission-type photocathode with the new laser focusing system.

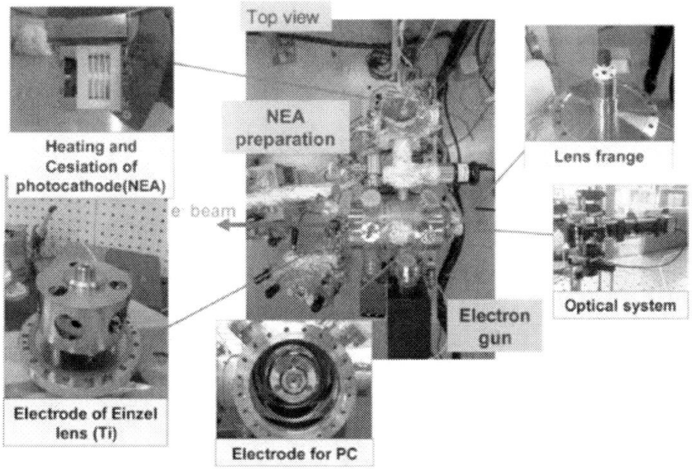

Figure 3. The photographs of the electron gun including an optical system and collector lens.

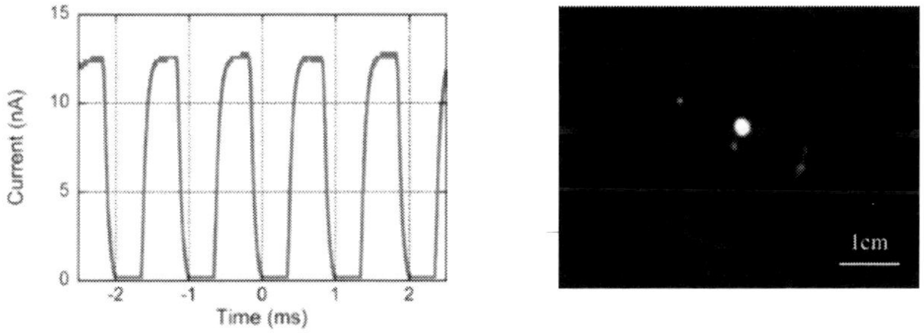

Figure 4. The left figure shows temporal current profile of the polarized electron beam extracted from the transmission-type photocathode under operation beam energy of 40keV. The right figure shows a phosphorous screen image of spatial beam profile

3.2. Photocathode

Requirements for the photocathode are (1) high ESP and (2) high QE. The latter is assured by a clean GaAs surface layer with high NEA. Both were already satisfied by a GaAs-GaAsP strained SL layer photocathode. However the previous photocathode is not compatible with the laser backward illumination, because GaAs was used as a substrate. Therefore, the micro-structure of the reflection type photocathode is modified to the transmission type. It was fabricated by MOVPE in Nagoya-U. The active layer is composed of 12 pairs of GaAs-GaAsP strained supperlattice (total thickness; 96 nm) with Zn-dopant concentration of 1.5×10^{18} cm^{-3}. The active-layer is grown on a 1-μm GaAsP buffer-layer which is deposited on a 500-nm GaAs inter-layer. These micro-film-structures are grown on a GaP wafer which works as the transparent substrate. The active-layer is coated with a highly p-doped GaAs surface-layer with a Zn concentration of 6×10^{19} cm^{-3}. The GaAs inter-layer plays an important role to achieve the highest ESP of ~90% by controlling the strain property of the GaAsP buffer-layer [14]. In the other 20-kV DC-gun, the photocathode could produce a polarized electron beam of the reduced brightness of $1\text{x}10^{7}$ A $m^{-2}sr^{-1}V^{-1}$. It is noted that the brightness was dramatically improved, that is 1000 times larger than that of conventional PES [15]. The pulse performance of the

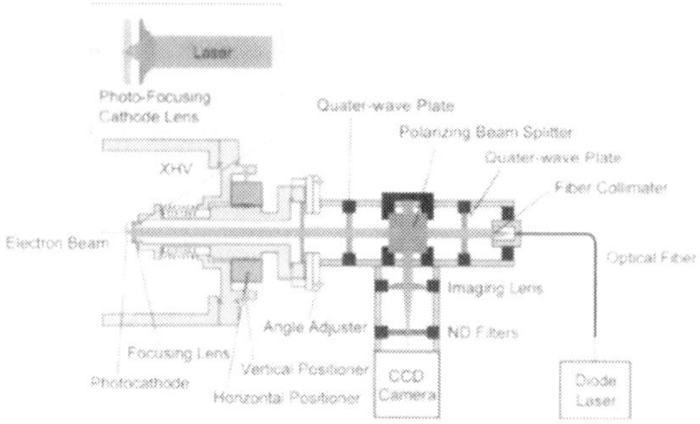

Figure 5. Schematic of the optical system to create and drive the spin-polarized electron beam.

PES in creation of a sub-nanosecond bunch-beam was also realized by using a high p-doped surface that can relax a surface charge limit phenomena [11].

3.3. Laser illumination system

Figure 5 shows a schematic of the laser optical system. To make a compact optical system, a cw light emitted from a semiconductor laser diode, which has linear polarization and wavelength of 785 nm, is transferred to the PES through a polarization-maintaining fiber. The linearly polarized light injected from the fiber is converted to circularly polarized light by a quarter-wave plate that can switch the light helicity from +1 to -1. The circularly polarized light is then guided and focused on an active SL layer of the photocathode by an aspheric lens. The position of the lens is adjusted by monitoring the reflected laser spot size by projecting a magnified focus image on a charge coupled device (CCD) camera. Using the transmission type photocathode, the laser spot size could be minimized to a radius of 0.65 μm in half width at half maximum, which is the diffraction-limit of 785 nm wavelength due to the NA of the focusing lens.

4. Spin manipulation

4.1. Spin rotation in longitudinal axis

A spin-rotator is necessary to rotate a spin direction of polarized electron beam. The design of the SPTEM system also requires a 90 degrees deflection of electron beam because of the angle between a beam axis of the electron gun and that of lens barrels in the image-projection optics. Then the instrument has to realize both spin-rotation and beam deflection. An arbitrary angle rotation of the polarization is achieved by the combination of crossed electric and magnetic fields whose proportion defines the polarization angle. The angle between the beam direction and the spin direction can be controlled to any polar angles. The spin can be rotated in transverse plane azimuthally by a field of SRL magnet in the series of CL magnets. The spin rotation apparatus was designed carefully to maintain both of UHV and high brightness conditions. An electric field is created by a sectored spherical condenser with the curvature radius of 100 mm. A transverse magnetic field is generated by a magnetic pole pair with the gap separation of 30 mm. In order to avoid a degradation of UHV, magnet coils to provide the magnetic field is located outside of the vacuum chamber, and the magnetic field is lead through the magnetic poles into the vacuum.

4.2. CL magnets and spin rotation in-plane

A CL system essentially exists for collimating and concentrating the beam to a specimen. In this new system, the CL system also plays a role of spin-rotator for in-plane rotation. The system consists of four lens magnets, which are SCL1, SRL, SCL2 and SCL3. In beam optics, the electron beam extracting from the 90-degrees bending-type spin-rotator is focused in the center of SRL by the SCL1 solenoid magnet, since the beam trajectory does not have to be disarranged in the SCL2 regardless of any magnitude of the magnetic field in the SRL. The SCL2 and SCL3 play the role of an ordinary CL in TEM. Therefore the newly designed CL system can realize an in-plane spin-rotation with concentrating the beam to a specimen.

5. Summary

The 40keV PES for SPTEM has been constructed and performed with a transmission-type photocathode. The 90-degrees bending spin-rotator and the illuminating-lens system including a CL system within in-plane spin-rotation have been made. The photocathode type PES has a high potential to observe a dynamics mechanics in a nano-structural matters with spin information [13, 17].

Acknowledgments

This research is supported by Grants-in-Aid for Scientific Research (S) (No. 51996964) and Grants-in-Aid for Young Scientists (B) (No. 22740157) from the MEXT of Japan, a Research fund of KEK for cooperative development of polarized electron sources and Technology Development Program for Advanced Measurement and Analysis of Japanese Science and Technology Agency (Group Leader; T. Nakanishi).

References

[1] Prescott CY, Atwood WB, Cottrell RLA, Destaebler H, Garwin EL, Gonidec A, Miller RH, Rochester LS, Sata T, Sherden DJ, Sinclair CK, Stein S, Taylor RE, Clendenin JE, Hughes VW, Sasao N, Schuler KP, Borghini MG, Lubelsmeyer K, Jentschke W 1978 *Phys. Lett.* B **77** 347
[2] SLD Collaboration 1995 *Phys. Rev. Lett.* **74** 2880
[3] Zdyb R and Bauer E 2002 *Phys.Rev. Lett.* **88** 166403
[4] Kohashi T and Koike K 2001 *Jpn. J. Appl. Phys.* **40** 1264
[5] N. Gisin, G. Ribordy, W. Tittel, and H. Zbinden 2002 *Rev. Mod. Phys.* **74**, 145
[6] Loss D and DiVincenzo DP, 1998 *Phys. Rev.* A **57** 120
[7] Kuwahara M, Kutsuwa T, Ono K, Kosaka H 2001 *Appl. Phys. Lett.* **96** 163107
[8] Tanaka N, Nakanishi T, Takeda Y, Asano H, Saitoh K, Ujihara T and Kuwahara M 2010 *Proceedings of 17th International Microscopy Congress* (Rio de Janeilo) I20.2
[9] ILC collaboration 2007, *ILC Global Design Effort and World Wide Study*
[10] Nishitani T, Nakanishi T, Yamamoto M, Okumi S, Furuta F, Miyamoto M, Kuwahara M, Yamamoto N, Naniwa K, Watanabe O, Takeda Y, Kobayakawa H, Takashima Y, Horinaka H, Matsuyama T, Togawa K, Saka T, Tawada M, Omori T, Kurihara Y, Yoshioka M, Kato K and Baba T 2005 *J. Appl. Phys.* **97** 094907
[11] Togawa K, Nakanishi T, Baba T, Furuta F, Horinaka H, Kurihara Y, Matsumoto H, Matsuyama T, Nishitani T, Okumi S, Omori T, Suzuki C, Takeuchi K, Wada K, Yamamoto M and Yoshioka M 2000 *Nucl. Instrum. Methods Phys. Res.* A **455** 118
[12] Furuta F, Nakanishi T, Okumi S, Gotou T, Yamamoto M, Miyamoto M, Kuwahara M, Yamamoto M, Naniwa K, Yasui K, Matsumoto H, Yoshioka M and Togawa K 2005 *Nucl. Instr. and Meth.* A **538** 33
[13] Kato S, Aono M, Sato K and Baba Y 1990 *J. Vac. Sci. Technol* A **8** 2860 (1990)
[14] Jin X, Yamamoto N, Nakagawa Y, Mano A, Kato T, Tanioku M, Ujihara T, Takeda Y, Okumi S, Yamamoto M, Nakanishi T, Saka T, Horinaka H, Kato T, Yasue T and Koshikawa T 2008 *Appl. Phys. Exp.* **1** 045002.
[15] Yamamoto N, Nakanishi T, Mano A, Nakagawa Y, Okumi S, Yamamoto M, Konomi T, Jin X, Ujihara T, Takeda Y, Ohshima T, Saka T, Kato T, Horinaka H, Yasue T, Koshikawa T, Kuwahara M 2008 *J. Appl. Phys.* **103** 064905
[16] King WE, Campbell GH, Frank A, Reed B, Schmerge JF, Siwick BJ, Stuart BC, Weber PM 2005 *J. Appl. Phys.* **97** 111101
[17] Flannigan DJ, Park ST, Zewail AH 2010 *Nano Lett.* **10** 4767

Status of the high brightness polarized electron source using transmission photocathode

N. Yamamoto[1], X.G. Jin[1], A. Mano[1], T. Ujihara[1], Y. Takeda[1], S. Okumi[2], T. Nakanishi[2], T. Yasue[3], T. Koshikawa[3], T. Ohshima[4], T. Saka[5], and H. Horinaka[6]

[1]Graduate School of Engineering, Nagoya University, Nagoya, Japan
[2]Graduate School of Science, Nagoya University, Nagoya, Japan
[3]Osaka Electro-Communication University, Osaka, Japan
[4]Central Research Laboratory, Hitachi Ltd., Tokyo, Japan
[5]Daido Institute of Technology, Nagoya, Japan
[6]Faculty of Engineering, Osaka Prefecture University, Osaka, Japan

E-mail: naoto@nagoya-u.jp

Abstract. Recently, we have developed a transmission-type polarized electron sources (PES)s for the generation of a high brightness beam. The developed PES can applied the extraction voltage of 20 kV at 4 mm electrode-gap and enables to realize the source beam radius of a few micro meter. As the results of prototype gun experiments, the brightness of $\sim 2 \times 10^7$ $A.cm^{-2}.sr^{-1}$ and a charge density lifetime of 1.8×10^8 $C.cm^{-2}$ were obtained. A maximum polarization of ~90 % and quantum efficiency of 0.09 % was achieved simultaneously using the transmission photocathode with GaAs-GaAsP strained superlattice layers. Up to now, another two electron gun was manufactured, and those mechanical designs are almost same as that of the prototype. The prototype gun and the second gun have been already operated for the photocathode R&D and for the Spin-LEEM application, respectively. In the near future, further experiments are prepared by using the third gun.

1. Introduction

In order to produce a high brightness beam with a high electron spin polarization (ESP), we have developed transmission-type polarized electron source (PES)s [1, 2]. The schematic view of the PES is shown in Fig.1. The PES consists of the transmission photocathode with GaAs-GaAsP superlattice layers and a 20-kV DC-gun with μm laser focusing system.

In the reference papers [1, 2], we reported the specifications of the prototype PES named "JPES-1". With regard to the photocathode, the GaAs-GaAsP strained strained superlatice was grown on GaP substrate, which is transparent for the laser of around 800 nm wavelength, and an ESP of ~90 % and quantum efficiency (QE) of 0.09 % were achieved simultaneously at 20 kV extracted voltage. Concerning the 20-kV DC-gun, a laser spot diameter of a few μm was obtained on the photocahode surface and the brightness of $\sim 2 \times 10^7$ $A.cm^{-2}.sr^{-1}$ was achieved. These performances are almost equal in ESP and better in brightness by three orders of magnitude, compared with those achieved by the no-transmission PES [3, 4].

Encouraged by these results, another two electron gun named "JPES-2" and "JPES-3" was manufactured by the same mechanical designs as that of JPES-1. JPES-1 has been used for the

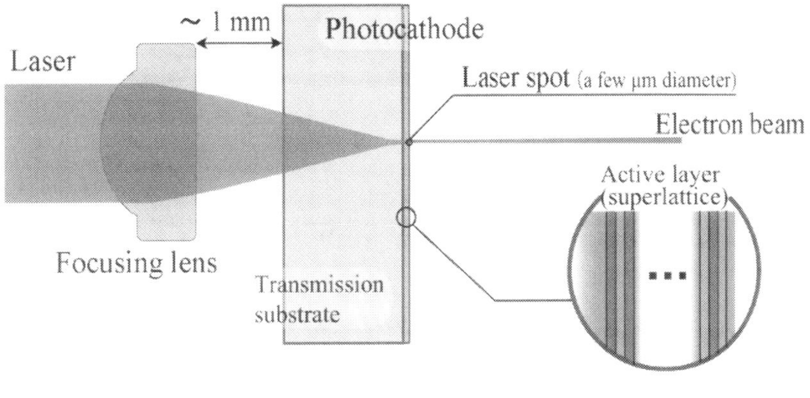

Figure 1. Schematic view of the Transmission PES

photocathode R&D at Nagoya University. JPES-2 had been already installed in the Low Energy Electron Microscope (LEEM) apparatus operated at Osaka Electro-Communication University (OECU). In the Spin-LEEM experiments at OECU, we had demonstrated a dynamic observation of magnetic images during the growth of Co on W(110) with an acquisition time of 0.02 s with a field view of 6 μm [5]. The acquisition time of a magnetic image by the conventional Spin-LEEM is between fractions of seconds and several tens seconds, which depends on the degree of the magnetic contrast, and the acquisition time of 0.2 s/frame without switching of spin direction has been performed [6]. Furthermore, several experiments by using JPES-3 are also scheduled now and the details will be described elsewhere.

In this paper, we reviewed the recent results of the photocathode R&D and the operation status of electron guns.

2. Transmission Photocathode

The transmission photocathodes were fabricated by using a low-pressure organometallic vapor phase epitaxy (OMVPE) system with a vertical cold-wall quartz reactor at Nagoya University. A commercially available GaP wafer with a Zn dopant of 1.4×10^{17} cm^{-3} was used as a transmission substrate. To control the strain property in the buffer and SL layers [2] [7], the GaAs inter-layer was grown on the GaP substrate. Then, 1 μm of the GaAsP buffer layer and 12 pairs of SL layers were grown with a Zn dopant concentration of 1.5×10^{18} cm^{-3}.

By using this photocathode and JPES-1, an ESP of 90 % and QE of 0.09 % was achieved simultaneously at 20 kV extracted voltage. The perfomances are almost equal in ESP but a bit lower in QE, compared with those achieved by the conventional photocathode [3, 4]. However, we believe that the polarization improved as a result of the control over strain relaxation in the buffer layer. Because the absorption of the laser power by the GaAs inter-layer was estimated about 50 % and we observed many crosshatches in an atomic force microscope image, which indicates the relaxation of tensile strain in the GaAsP layer [8]. We are doing the improvement designs of the photocathode and QE measurements now.

3. 20–kV electron gun

3.1. Electrode design and dark current measurement

Three electron guns were developed at Nagoya University and the accelerating electrodes were designed by using the own simulation code [9] of N.Y.. At first of JPES-1 experiments, a pair of titanium electrodes was installed at Gun chamber and the accelerating voltage and the electrode gap were set to be 20 kV and 5 mm respectively. In the experiment of these electrodes, dark currents of a few nA were observed at 20 kV. Since the electrode made of a molybdenum cathode and titanium anode was found to suppress dark currents by our previous study [10], we replaced the titanium cathode to the molybdenum one and the electrode gap were also changed to be 4 mm. As a result, dark currents were successfully suppressed to at most 1 nA at 20 kV and a few nA at 30 kV. Details of these experiments will be reported in another paper.

3.2. Beam brightness measurement

For the purpose to calculate the beam brightness, the beam profile was measured using the knife-edge method at 53.1 cm downstream from the photocathode. The angular current density $dI/d\Omega$ and brightness B of the electron beam are calculated by the following equations[1],

$$\frac{dI}{d\Omega} = I \cdot \frac{L^2}{\pi(S_L - S_o)^2}, \quad (1)$$

$$B = \frac{1}{\pi \cdot S_o^2} \cdot \frac{dI}{d\Omega}. \quad (2)$$

Here, L is the drift length between the source and the above brightness monitor (BM), and S_o is a virtual source radius at the photocathode surface. S_L and I represent the beam radius and current measured by the BM. For S_o, the virtual source size must be employed due to the beam focusing property caused by the electrode design, and it was estimated to be about 70% smaller than the emission area by beam tracking simulations.

The laser spot size could be well measured by a charged coupled device, but the emission area was enlarged due to the diffusion effect inside the photocathode. We assume that the diffusion length in the photocathode is 0.25 μm.

The data was taken under a emission current of 5.3 μA with a laser spot radius of 0.65 μm and an angular current density of 248 mA/sr and a brightness of $2.0 \pm 0.8 \times 10^7$ A.cm^{-2}.sr^{-1} were obtained. This brightness is better by three orders of magnitude, compared for the conventional PES.

3.3. Vacuum condition and beam lifetime

The NEA surface is known to be easily degraded by the desorption of residual gas molecules and the back bombardment of positive ion produced by the beam itself. Therefore the base pressure near the photocathode was maintained less than 1.0×10^9 Pa by using an 100 L/s ion pump and a 2000 L/s non-evaporable getter pump.

The QE lifetimes without electron beams were measured at JPES-1. Two series of data for different conditions were shown in Fig. 2. The open circle points (red color) indicate the results with the condition of no laser and extracted voltage of 20 kV, and the open square (blue) with the estimated laser density of 473 W/mm^2 on the photocathode and no extracted voltage. The vacuum pressure at measurements was 1.4×10^{-9} Pa and the dark currents between electrodes were less then 1 nA at 20 kV extracted voltage. Both results seems to be constant, demonstrating that there is no observable evidence of the QE degradation due to dark currents and laser heating.

We have also measured beam lifetimes during the experiments of JPES-1. The measurements were carried out with beam current decay mode and the extracted voltage of 20 kV. The lifetime was defined as the period until the QE value decreased to 1/e of the initial value. The charge density lifetimes as a function of the operation pressure are shown in Fig. 3 and three typical data of the Fig. 3 was summarized in Tab. 1. The charge density lifetimes increase as the operation pressures decrease and a maximum value of 1.8×10^8 C.cm^{-2}was achieved. The data was obtained with the initial current of 3 μA and the laser radius of 3.2 μm, corresponding to the beam lifetime of 36 hours and the charge lifetime of 260 mC (See Tab. 1). Especially the

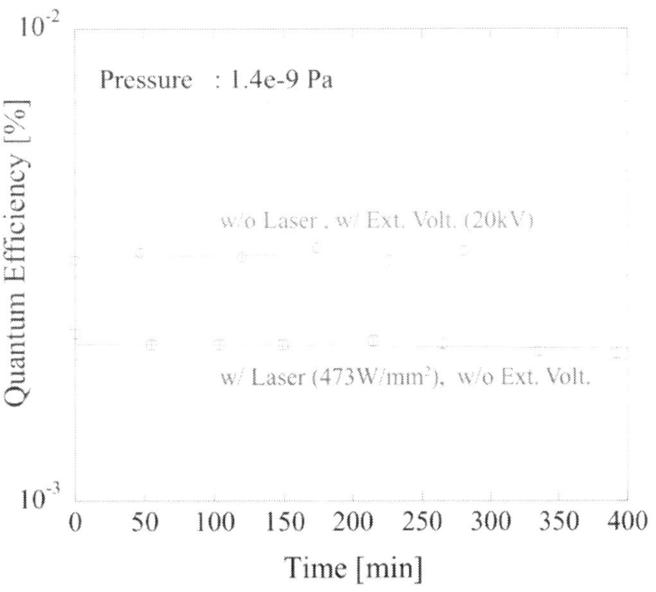

Figure 2. Dark lifetime with two different conditions

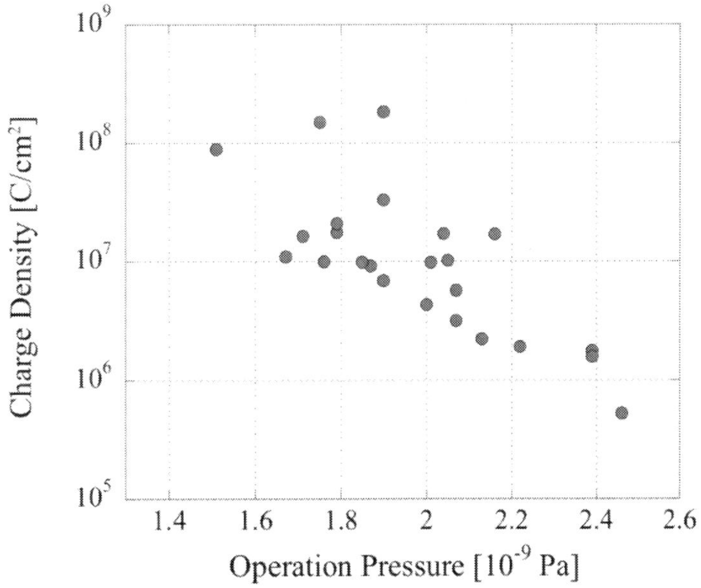

Figure 3. Charge density lifetime as a function of operation pressure

operation pressure during SPLEEM experiments, which needs the beam currents of around 1 μA, were less than 1×10^{-9}. Then reactivation of a photocathode surface is not necessary for two or three months.

Initial Current (μA)	Current Density (mA/mm^2)	$1/e$ lifetime (hour)	Total charge (mC)	Charge Density (C/cm^2)
3	82	36	260	1.8×10^8
2	55	50	213	1.5×10^8
1	27	55	126	8.8×10^7

Table 1. Typical data of beam lifetime measurements

4. Summary

The high brightness PESs were developed by using transmission-type photocathodes with GaAs-GaAsP strained superlattice layers and electron guns with μm laser focusing system. The developments were successfully done and a brightness of $2.0 \pm 0.8 \times 10^7$ A cm^{-2} sr^{-1} and an ESP of 90% were simultaneously achieved at 20 kV beams. Furthermore very highly charge density lifetime are obtained with the extreme high vacuum conditions. Up to now, three electron guns are manufactured at Nagoya University, and those are used for photocathode R&Ds and applications of polarized electron beam.

At present except for the QE value, the performances of transmission photocathodes are higher compared with conventional photocathodes and the photocathodes could be applied not only for material science but also high energy accelerators, such as a linear collider and an energy recovery linac. To demonstrate promising advantages of transmission photocathode for these applications, further experiments are required. Then we are prepared for the generation of the multi-bunch beam and the emittance measurement in near future.

Acknowledgments

The authors wish to thank the SPLEEM collaboration for their help during the experiments. One of the authors (N. Y.) has been financially supported by a Grant-in-Aid for Scientific Research (B21740175) and the Tatematsu Foundation.

References

[1] Yamamoto N, Nakanishi T, Mano A, Nakagawa Y, Okumi S, Yamamoto M, Konomi T, Jin X, Ujihara T, Takeda Y, Ohshima T, Saka T, Kato T, Horinaka H, Yasue T, Koshikawa T and Kuwahara M 2008 *Journal of Applied Physics* **103** 064905

[2] Jin X, Yamamoto N, Nakagawa Y, Mano A, Kato T, Tanioku M, Ujihara T, Takeda Y, Okumi S, Yamamoto M, Nakanishi T, Saka T, Horinaka H, Kato T, Yasue T and Koshikawa T 2008 *Applied Physics Express* **1** 045002

[3] Nakanishi T, Togawa K, Baba T, Furuta F, Horinaka H, Kato T, Kurihara Y, Matsumoto H, Matsuyama T, Nishitani T, Okumi S, Omori T, Saka T, Suzuki C, Takeuchi Y, Wada K, Wada K, Yamamoto M and Yoshioka M 2000 *Nuclear Inst. and Methods in Physics Research, A* **455** 109

[4] Nishitani T, Nakanishi T, Yamamoto M, Okumi S, Furuta F, Miyamoto M, Kuwahara M, Yamamoto N, Naniwa K, Watanabe O, Takeda Y, Kobayakawa H, Takashima Y, Horinaka H, Matsuyama T, Togawa K, Saka T, Tawada M, Omori T, Kurihara Y, Yoshioka M, Kato T and Baba T 2005 *Journal of Applied Physics* **97** 94907–94907

[5] Suzuki M, Hashimoto M, Yasue T, Koshikawa T, Nakagawa Y, Konomi T, Mano A, Yamamoto N, Kuwahara M, Yamamoto M *et al.* 2010 *Applied Physics Express* **3** 026601

[6] Ramchal R, Schmid A K, Farle M and Poppa H 2003 *Phys. Rev. B* **68** 054418

[7] Jin X, Maeda Y, Saka T, Tanioku M, Fuchi S, Ujihara T, Takeda Y, Yamamoto N, Nakagawa Y, Mano A, Okumi S, Yamamoto M, Nakanishi T, Horinaka H, Kato T, Yasue T and Koshikawa T 2008 *Journal of Crystal Growth* **310** 5039 – 5043 ISSN 0022-0248 the Fourteenth International conference on Metalorganic Vapor Phase Epitax, The 14th International conference on Metalorganic Vapor Phase Epitax

[8] Jin X, Maeda Y, Sasaki T, Arai S, Ishida Y, Kanda M, Fuchi S, Ujihara T, Saka T and Takeda Y 2010 *Journal of Applied Physics* **108** 094509 (pages 6)

[9] Yamamoto N, Mizuno A, Masuda K, Kan K, Matsumoto S and Yamamoto M 2009 *AIP Conference Proceeding* vol 1149 ed Crabb D G, Prok Y, Poelker M, Liuti S, Day D B and Zheng X (AIP) pp 1174–1178

[10] Furuta F, Nakanishi T, Okumi S, Gotou T, Yamamoto M, Miyamoto M, Kuwahara M, Yamamoto N, Naniwa K, Yasui K, Matsumoto H, Yoshioka M and Togawa K 2005 *Nuclear Instruments and Methods in Physics Research, Section A: Accelerators, Spectrometers, Detectors and Associated Equipment* **538** 33–44

Measurement of Electron Beam Polarization from Unstrained Bulk GaAs via Two Photon Photoemission

J L McCarter[1], T J Gay[2], J Hansknecht[3], M Poelker[3], M L Stutzman[3]
[1] Department of Physics, University of Virginia, Charlottesville, VA 22901
[2] Behlen Laboratory of Physics, University of Nebraska Lincoln, Lincoln, NE, 68588
[3] Thomas Jefferson National Accelerator Facility, 12050 Jefferson Ave. Suite 500, Newport News, VA 23606

Email: [1]jlm2ar@virginia.edu

Abstract. This paper describes measurements of the beam polarization and quantum efficiency for photoemission using two-photon excitation from unstrained bulk GaAs illuminated with pulsed, high intensity 1560nm laser light. Quantum efficiency is linearly proportional to 1560nm peak laser intensity, which was varied in three independent ways, indicating that the emitted electrons are promoted from the valence to the conduction band via two-photon absorption. Beam polarization was measured using a microMott polarimeter, with a value of 16.8(4)% polarization at 1560nm, which is roughly half the measured value of 33.4(8)% using 778 nm light.

1. Introduction

Polarized electron sources are important components of particle accelerators like the Continuous Electron Beam Accelerator Facility (CEBAF) at Jefferson Lab, where the polarized electron beam is used to study nuclear structure, the dynamics of strong interactions, electro-weak nuclear physics including parity-violation, and physics beyond the Standard Model. The first GaAs-based polarized electron source used at an accelerator [1] provided beam polarization ~ 35%, with polarization limited to less than 50% by the energy level degeneracy in the $^2P_{3/2}$ valence band state of unstrained bulk GaAs (Figure 1a) [2]. Significantly higher beam polarization was obtained in the 1990s by introducing an axial strain within the GaAs crystal structure to eliminate this degeneracy [3,4]. Today, beam polarization routinely exceeds 80% using strained-superlattice GaAs/GaAsP structures [5,6,7,8]. However, these high polarization photocathodes are expensive and delicate compared to unstrained bulk GaAs, making them more difficult to use.

This paper presents a summary of measurements of quantum efficiency and polarization of photo-emitted electrons excited through two-photon absorption from unstrained bulk GaAs, as analyzed by a compact retarding-field microMott polarimeter.

2. Theoretical background

Nakanishi first proposed using two-photon absorption with unstrained bulk GaAs to obtain high polarization [9]. He reasoned that quantum mechanical selection rules associated with the absorption of two photons of circularly-polarized light at half of the bandgap energy would allow excitation of only one spin state to the conduction band (Figure 1b). Although the efficiency of this non-linear optical process would be small, the observation of beam polarization > 50% from unstrained bulk

GaAs would be an exciting development, giving insight into the band structure and allowed quantum mechanical transitions in GaAs, and would be useful in a practical sense because research projects with polarized electrons often require only very low current beam. Furthermore, recent advances in the lasers used for telecommunications have made high power light sources at 1560nm inexpensive and abundant. Photoluminescence experiments support this theory with polarization measured at 58% and extrapolated to infer that the polarization at excitation was 95% [10].

A more sophisticated theory by Bhat et al. [11] suggests electrons make transitions from the valence and conduction bands via two-photon absorption as governed by quantum mechanical selection rules (as suggested by Nakanishi), but also via "forbidden" intraband transitions within conduction or valence bands, with angular momentum of the photons transferred to carrier orbital motion and the crystal structure, and not just the electron spin. They predict polarization via two-photon absorption will be comparable to conventional one-photon induced polarization, a value less than 50%. Experimental results supporting this theory used differential transmission pump/probe techniques to measure nascent polarization equal to 48% [11,12].

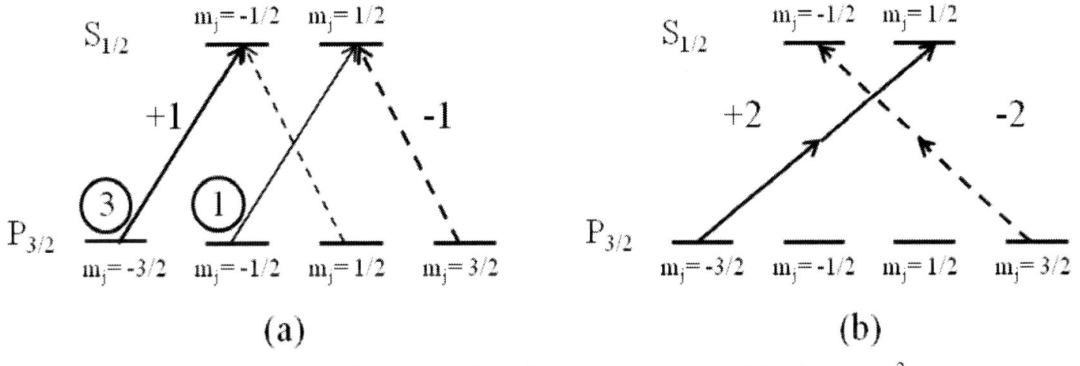

Figure 1. Optical transitions in GaAs with circularly polarized light from the $^2P_{3/2}$ valence band to the $^2S_{1/2}$ conduction band: a) one-photon excitation, and b) two-photon excitation, as suggested by Nakanishi. The circled values indicate transition probabilities. Maximum polarization from un-strained bulk GaAs would be 50% for one photon excitation and 100% for two photon excitation.

Two-photon absorption is a non-linear optical process [13] that only occurs within crystals that lack inversion symmetry, such as GaAs. The cross-section for this process is small, meaning the likelihood that two photons with energy less than the band gap can be absorbed simultaneously to excite an electron from the valence band to the conduction band is small compared to conventional one-photon absorption with light having more energy than the band gap.

For conventional, one-photon photoexcitation, the number of electrons emitted is proportional to the number of incident photons, or laser power, and is constant with laser intensity. Equation 1 [11, 14,15] relates the expected QE at 1560 nm to that at 778 nm:

$$\frac{QE_{1560}}{QE_{778}}=4\pi e^{\frac{2}{hc}}\frac{(2h\omega-E_{gap})}{h\omega}\frac{I}{m_c\omega^3}=2*10^{-12}*I(\frac{W}{cm^2}) , \tag{1}$$

where $\omega=\frac{2\pi c}{\lambda}$ is the frequency of the light, I is the light intensity, m_c is the electronic effective mass, and E_{gap} is the band gap energy.

There are two noteworthy features of this equation. First, QE_{1560nm} due to two-photon absorption is proportional to laser intensity, and secondly, QE_{1560nm} will be very small unless intensity is quite large. There are three relatively simple ways to provide high intensity for this experiment: a) illuminate the photocathode with high average power, b) use short-pulse light to generate high peak power, and c) focus the incident laser beam to a small spot at the photocathode. Laser intensity was varied through each of these methods to verify the two-photon absorption and photoemission process.

3. Experiment

A load-locked vacuum apparatus was employed, consisting of three chambers (Figure 2a): a low-voltage source chamber for installing and activating photocathodes, a beam transport section, and a microMott retarding-field polarimeter. Un-strained bulk GaAs was mounted to a long stalk that could be lowered into the source chamber and activated to negative electron affinity (NEA) with Cs and NF_3. The photocathode was biased at -268V and the emitted electron beam was delivered to the microMott polarimeter using a 90° electrostatic deflector [16] and electrostatic steering lenses [17]. The beam transport system and the microMott polarimeter are described thoroughly in another publication [18]. Optical systems for each wavelength, 778nm and 1560nm, could be quickly and reproducibly moved in and out of position beneath the vacuum chamber. They comprised numerous elements manufactured for the appropriate wavelength: optical attenuators, linear polarizers, quarter-wave plates, and insertable half-wave plates. When the 1560nm laser system was employed, long pass optical filters (two at 1350nm and one at 850nm) were inserted to ensure that no light below 850nm could enter the vacuum apparatus. This was very important, because fiber amplifiers (described below) produce a small amount of light in the visible and near-IR wavelengths via second harmonic generation. Even a small amount of undetected light in the visible or near-IR wavelength range could easily overwhelm the beam produced at 1560nm where two-photon photocurrent is comparatively small. Insertion of additional 1350nm long pass filters had no effect on the amount of photocurrent, indicating that only 1560nm light reached the photocathode. Finally, a mirror system mounted to dual translation stages directed the laser beam vertically into the source chamber and provided a means to "QE scan" the photocathode (see below).

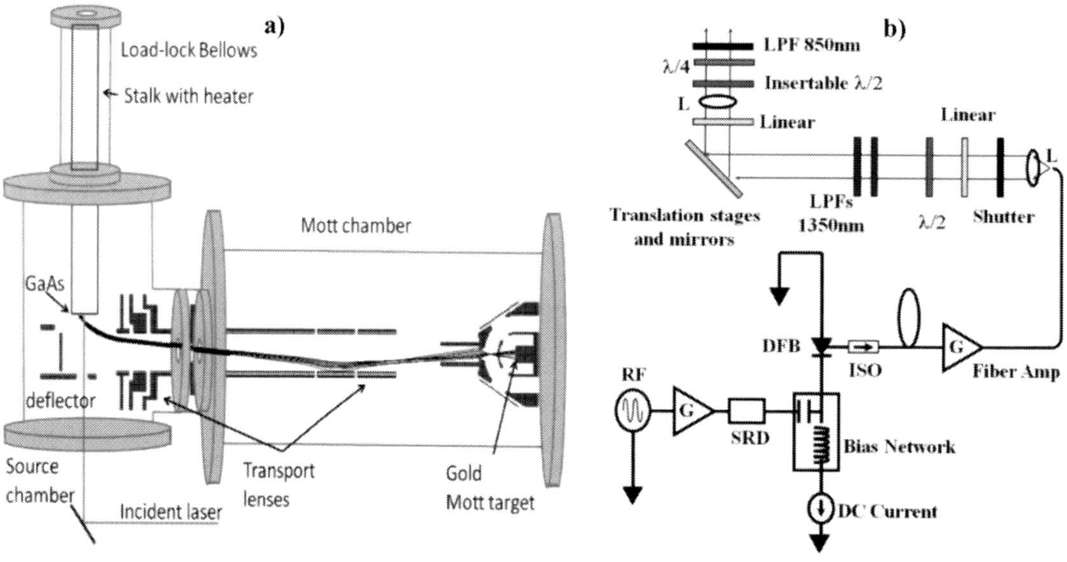

Figure 2. a) Schematic of the experimental apparatus, b) Schematic of 1560nm light source used to generate two-photon photo-emitted beam from un-strained bulk GaAs. DFB, distributed feedback Bragg reflector diode laser; ISO, fiber isolator; SRD, step recovery diode; L, lens; Linear, linear polarizer; λ/2 and λ/4, half-wave and quarter-wave plates; LPF, long pass filter.

The 778nm laser was a simple low-power diode laser that could be operated in DC or RF pulsed-mode via gain-switching at repetition rates from 250 to 2000 MHz. The 1560nm laser was composed of a gain-switched fiber-coupled diode "seed" laser and fiber-amplifier (Figure 2b), that could produce up to 5W average power with 35ps optical pulses over the same range of pulse repetition rates as the 778nm laser system [19]. By using short pulse length light, high peak power was obtained to enhance the two-photon absorption process.

As mentioned above, the QE_{778nm} for clean and undamaged GaAs should be independent of incident laser power, intensity, peak power or laser spot size. For two-photon absorption, QE_{1560nm} should vary linearly with laser intensity. In this experiment, there were three "knobs" to vary the intensity at the photocathode: the average power of the light, the spot size of the laser beam on the photocathode, and the pulse repetition rate (while keeping pulsewidth and average power constant). Equations 2 and 3 illustrate how these parameters affect intensity at the photocathode:

$$I_{avg} = P_{avg}/A \quad , \tag{2}$$

$$I_{peak} = \frac{I_{avg}}{D} = \frac{P_{avg}}{A*D} = \frac{P_{avg}}{A*t*f} \, , \tag{3}$$

where P_{avg} is the average power of the laser, D is duty factor equal to the product of the optical pulsewidth t and the laser pulse repetition rate f, and A is the spot size of the laser on the photocathode. High intensity was obtained by increasing the average power of the incident laser light using an optical attenuator, and by decreasing the size of the laser beam. Different laser spot sizes were obtained by simply using lenses of different focal length. A 30cm focal length lens produced a beam waist (i.e., the smallest spot) at the photocathode and longer focal length lenses placed in the same location below the vacuum window produced larger spots at the photocathode. Because changing lenses in the optical system can steer the beam, QE scans were done after each lens change (see below).

The rf-pulsed nature of the laser systems provided a means to easily vary the peak intensity of the light at the photocathode. The optical pulsewidth of the gain-switched laser system remained approximately constant across the range of pulse repetition rates: 35ps FWHM. The fiber amplifier provided the same output power for each repetition rate, so the peak intensity of the laser system could be varied by changing the pulse repetition rate, with the highest intensity obtained at lower pulse repetition rates.

The unstrained bulk GaAs wafer was activated, and reactivated as needed by heating to ~ 550°C to clean the surface and by applying Cs and NF_3 to make a NEA surface [20]. Each measurement comparing one and two-photon emission was taken during the same activation. Due to non-uniformity of the Cs deposition at the photocathode, the QE of the GaAs wafer was not constant over its surface. Where relevant, data are presented at the location of highest QE and at the location of the highest transmission to the polarimeter, which were not identical. The highest transmission - about 20% of the extracted beam delivered to the polarimeter - occurred from a specific location of the GaAs photocathode about 2mm in diameter. A QE scan of the photocathode was performed each time a laser system was installed, to identify the region of highest transmission and to ensure that comparative measurements between 778 and 1560nm laser systems were made using beam extracted from the same region of the photocathode.

4. Results

4.1 Two Photon QE

The unstrained bulk GaAs was first evaluated with 778nm light. As expected, QE_{778nm} remained nearly constant versus input power and pulse repetition rate (Figure 3a). In fact, there was a slight decrease in QE at higher laser power (made more apparent by plotting with zero suppression), observed for all repetition rates; this was attributed to the surface charge limit effect [21]. The QE behavior at 1560nm however, was markedly different (Figure 3b), as higher intensity produced higher QE, increasing linearly with power and intensity, indicative of a two-photon absorption process. As described above, average power (and average intensity) was varied by adjusting an optical attenuator, whereas peak intensity was varied by selecting different pulse repetition rates. It should be noted that high average power (≥2W) heated the photocathode, which significantly degraded its lifetime [22]. Cold nitrogen gas was applied to the back surface of the photocathode stalk to help maintain the QE during measurements. The slight roll-over of QE plots at high average power is likely a result of photocathode heating.

The repetition rate of the laser was also changed while keeping the average power constant, which according to equation 3 changed the peak laser intensity. As can be inferred from Figure 3b, the QE was linear with the laser intensity, which was again indicative of a two-photon process.

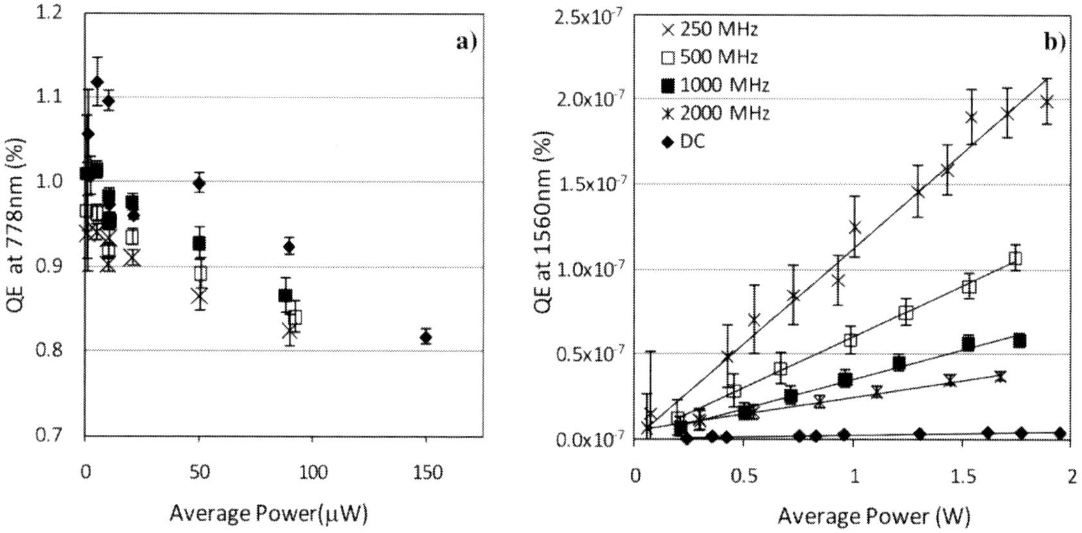

Figure 3. a) QE vs. input power at 778nm, adjusted by optical attenuator for a given repetition rate, and b) QE at 1560nm with the same repetition rates used in Figure 3a.

Further verification of two-photon photoemission was obtained by changing the size of the laser spot at the photocathode for both laser wavelengths, with results shown in Figure 4. The pulse repetition rate was 250MHz. As noted above, the photocathode activation was not uniform across the surface, so that the location of maximum QE was not coincident with the location of maximum beam transmission to the microMott polarimeter. The QE scans were performed to ensure that comparative measurements with both wavelengths were always made at the same photocathode location. Measurements in Figure 4 show results from both locations. At 778nm, the QE was nearly constant versus peak intensity, with maximum QE rising slightly at the smallest spot sizes due to the non-uniformity of the QE of the photocathode. The behavior at 1560nm however, was quite different from that at 780nm, namely, QE increases linearly with laser peak intensity (Figure 4b). Also shown is the theoretical predicted QE using a base QE at 778nm of 1.32%. The QE is proportional to the light intensity and is within an order-of-magnitude agreement with the predicted QE values, again indicating that two-photon absorption is the cause of the photocurrent.

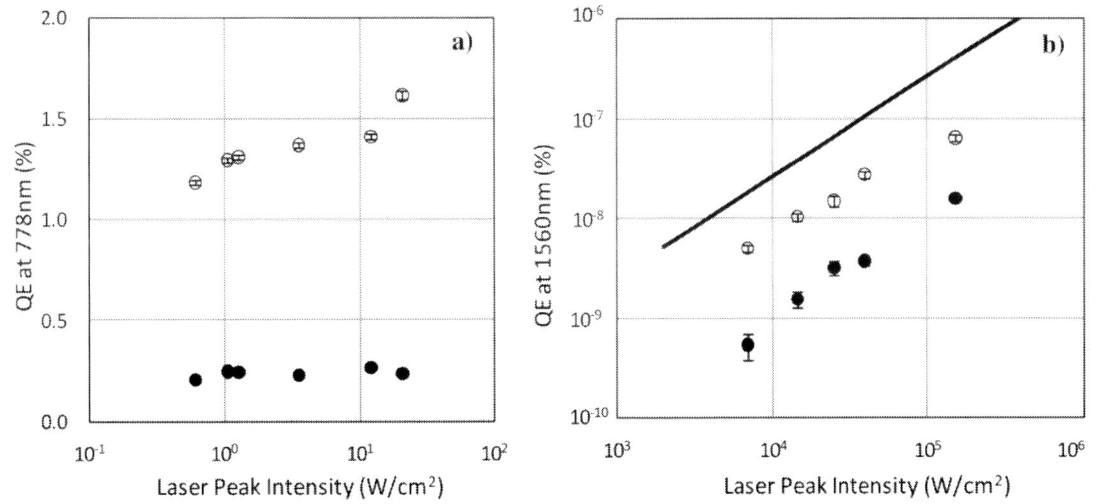

Figure 4. QE vs. peak intensity of the laser beam caused by changing the laser spot size at the cathode. The QE across the photocathode was not uniform: open circles represent maximum QE values, closed circles represent QE values at photocathode location for maximum transmission to the polarimeter. a) QE at 778nm. b) QE at 1560nm. The solid line is the theoretical QE (Eqn.1) for a two photon process at 1560nm, assuming a 1.3% QE at 778nm. Data were obtained using a laser repetition rate of 250MHz.

4.2 Two Photon Polarization

The polarization of electron beams generated with both one- and two-photon excitation was measured using the microMott polarimeter with its target biased at 20kV. For each polarization measurement, retarding field scans were performed to determine instrumental backgrounds. Repeated measurements, shown in Figure 5, indicate that the polarization for both wavelengths was stable with regard to extracted photocurrent. Photocurrents higher than 1nA were not tested using 1560nm light, as above this value the lifetime of the cathode was reduced to unusable levels due to laser heating of the GaAs. Measurements at 778nm were not made below 1nA in order to keep statistical errors low. Also included in Figure 5 is polarization data from a strained-superlattice GaAs photocathode taken using the same polarimeter, which indicates there is no instrumental dependence of the polarization on extracted photocurrent across the range of interest.

The effect of laser polarization on beam polarization is shown in Figure 6. The angle of the quarter-wave plate was rotated to vary the degree of circular polarization of the incident light. As expected, the electron polarization for both one- and two-photon excitation varies sinusoidally with the degree of circular polarization of the light. This behavior of the electron polarization with the degree of light polarization indicates that there is no issue with the optical arrangement that systematically affects the electron polarizations, which were measured to have maximum values of 16.8(4) % for 1560nm light and 33.4(8) % for 778nm light.

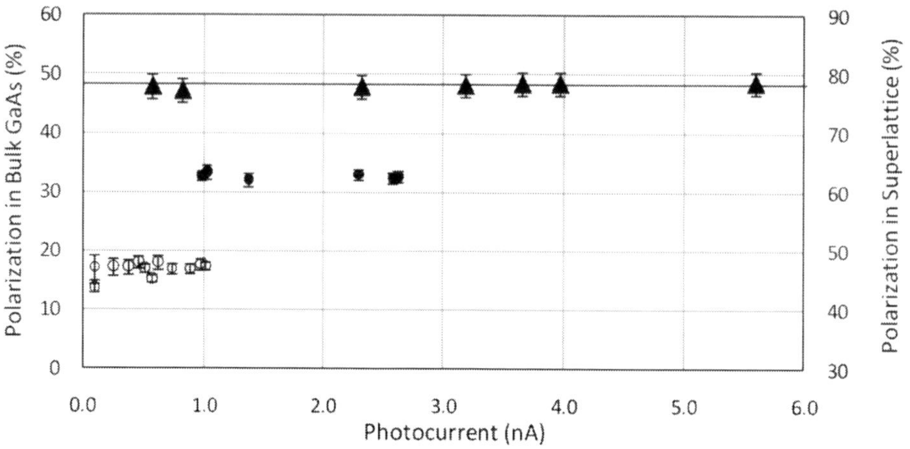

Figure 5. Polarization as a function of extracted photocurrent from unstrained bulk GaAs: 1560nm (open circles) and 778nm (closed circles); polarization measurements from strained-superlattice GaAs at 772nm (solid triangles), with the straight line representing the average value.

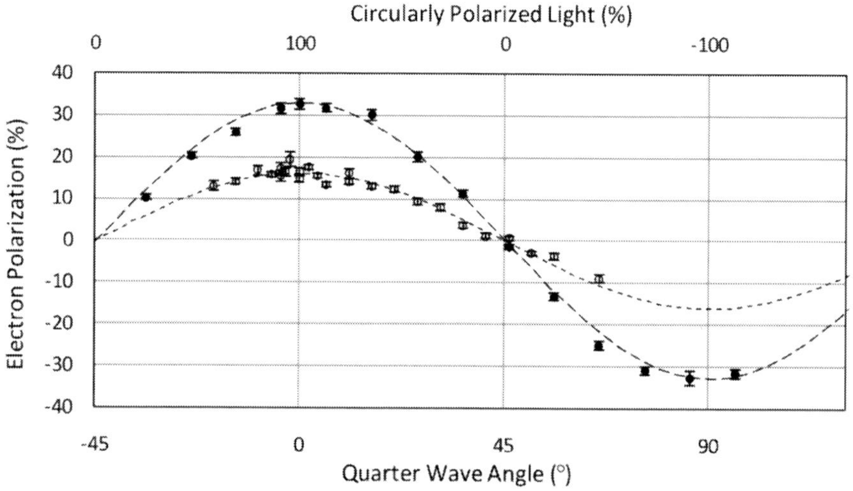

Figure 6. Polarization of the photoelectron beam at 778nm (closed circles) and at 1560nm (open circles) as a function of the orientation of the quarter-wave plate. Sinusoidal fits to the data are indicated by the dashed lines.

5. Conclusion

The polarization of electrons emitted by two-photon excitation of GaAs using 1560 nm laser light has been measured to be 16.8(4)%, roughly half the polarization using 778nm light, which is 33.4(8)%. The photoemission was confirmed to be due to two-photon absorption, as indicated by the linear dependence of the QE_{1560nm} on the incident laser intensity, varied using three different methods.

To explain this unexpected low polarization result, we speculate the electrons excited via two-photon absorption originate from relatively deep within the material, because the absorption coefficient is so small. Using a GaAs wafer polished on both sides, the absorption length for GaAs at 1560nm was measured to be ~220μm, while the absorption length at 778nm was ~1 μm. Measurements using time resolved polarimetry [23] show electrons depolarize in the bulk prior to emission from the surface. A larger proportion of electrons created via the two-photon process are created deeper inside the bulk material because of the attenuation lengths of the light used. The two-photon electrons have more time and distance to depolarize, thus decreasing the overall polarization of the photocurrent. Further studies are planned to confirm this hypothesis by using thin samples of bulk GaAs. Thinner GaAs samples should provide higher polarization, and it will be interesting to see if the two-photon polarization is always roughly half that of one-photon polarization.

Acknowledgements
Thanks to Steve Covert, Jim Clark and Phil Adderley for assistance with vacuum, mechanical, and electronics tasks. Thanks to Leonid Gerchikov of St. Petersburg State Polytechnic University, Russia for helpful discussions on understanding theoretical models of two-photon photoemission and polarization. Work supported by Jefferson Science Associates, LLC under U.S. DOE Contract No. DE-AC05-06OR23177 and by NSF Grant PHY-0821385.

References
[1] Prescott C Y, et al. 1978 *Phys. Lett.* B **77** 347-52
[2] J. Kessler 1985 *Polarized Electrons* (New York: Springer)
[3] Maruyama T, Garwin E L, Prepost R, Zapalac G H, Smith J S and Walker J D 1991 *Phys. Rev. Lett.* **66** 2376-9
[4] Nakanishi T, Aoyagi H, Horinaka H, Kamiya Y, Kato T, Nakamura S, Saka T and Tsubata M 1991 *Phys. Letters* A **158** 345-9
[5] Aulenbacher K, et al. 1997 *Nucl. Intrum. and Meth.* A **391** 498-506
[6] Von Drachenfels W, Frommberger F, Gowin M, Hillert W, Hoffmann M and Neff B 2003 *AIP Conf. Proc.* **675** 1053-7
[7] Zhou F, Brachmann A, Maruyama T and Sheppard J C 2009 *AIP Conf. Proc.* **1149** 992-6
[8] Poelker M 2009 *AIP Conf. Proc.* **1149** 174-9
[9] Matsuyama T, Suzuki S, Mukai M, Wada K, Horinaka H, Konda T, Hangyo M, Togawa K and Nakanishi T 2001 *4th Pacific Rim Conf. on Lasers and Electro-Optics* 164-5
[10] Matsuyama T, Horinaka H, Wada K, Kondo T, Hangyo M, Nakanishi T, Okumi S and Togawa K 2001 *Jpn. J. Appl. Phys.* **40** 555-7
[11] Bhat R D R, Nemec P, Kerachian Y, van Driel H M and Sipe J E 2005 *Phys. Rev.* B **71** 035209
[12] Miah M I 2009 *J. Phys. Chem.* B **113** 6800-2
[13] Yariv A 1985 *Optical Electronics* (New York: Holt, Rinehart and Winston)
[14] Ivchenko E L 1973 *Sov. Phys. Solid State* **14** 2942
[15] Arifzhanov S B and Ivchenko E L 1975 *Sov. Phys. Solid State* **17** 46
[16] Al-Khateeb H M, Birdsey B G, Bowen T C, Green A S, Johnston M E and Gay T J 1999 *Rev. Sci. Instrum.* **70** 3882
[17] Anderson T G, Birdsey B G, Woeher S M, Rosenberry M A and Gay T J 2001 *Rev. Sci. Instrum.* **72** 2923
[18] McCarter J L, Stutzman M L, Trantham K W, Anderson T G, Cook A M and Gay T J 2010 *Nucl. Instrum. and Meth.* A **618** 30-6
[19] Hansknecht J and Poelker M 2006 *PRSTAB* **9** 063501
[20] Sinclair C K, Adderly P A, Dunham B M, Hansknecht J C, Hartmann P, Poelker M, Price J S, Rutt P M, Schneider W J and Steigerwald M 2007 *PRSTAB* **10** 023501
[21] Woods M, Clendenin J, Frisch J, Kulikov A, Saez P, Schultz D, Turner J, Witte K and Zolotorev M 1993 *J. Appl. Phys.* **73** 8531-35
[22] Sinclair C K 2006 *Nucl. Instrum. and Meth.* A **557** 69-74
[23] Schuler J, et al. 1998 Time Resolved Polarization Measurements in the Photoemission of Strained Layer and Strained Superlattice Structures *Low Energy Polarized Electron Workshop 1998 Proceedings* (St. Peterburg, Russia, 2-5 September 1998) ed Y A Mamaev (Russia: SPES-Lab-Publishing) 20-23

Workshop on Sources of Polarized Leptons and High Brightness Electron Beams (PESP2010) IOP Publishing
Journal of Physics: Conference Series **298** (2011) 012019 doi:10.1088/1742-6596/298/1/012019

High average photocurrent research at MAMI

K. Aulenbacher, I. Alexander, E. Riehn, V. Tioukine

Institut für Kernphysik der Johannes Gutenberg-Universität zu Mainz, D-55099 Mainz

Abstract. We have measured the extractable charge during one lifetime (Charge lifetime, Q_τ) of a NEA-GaAs-photocathode for two beam diameters. A limitation of Q_τ to 1100 Coulomb due to ion-back-bombardment was observed for a 1.5 mm diameter emission site. When increasing the laser diameter from .3 to 1.5 mm we observe an increase of Q_τ by a factor ≈ 5, which is in disagreement with the assumed proportionality of the charge lifetime with emission area. Possible reasons for this discrepancy are analyzed.

1. Introduction

The purpose of this article is to present investigations which serve to demonstrate capabilities of photoemission sources for future projects, the demands of which are summarized in table 1.

Semiconductor cathodes with negative electron affinity (NEA-) surfaces are very efficient photoemitters, which even allow for spin-polarized photoemission. Unfortunately, the NEA state is a very delicate one, therefore the cathodes are unstable - they suffer from a reduction of quantum efficiency (q.e.) which is characterized by the time constant τ, the so-called 'lifetime'. For future large scale facilities as listed in table 1, lifetimes exceeding one month could be considered as acceptable.

Table 1. MAMI photoemission source performance compared to demands for future facilities. Charge and fluence lifetime (definitions see text) refer to a lifetime of one month.

Project	average current $I_b\,[A]$	av. current density $j_b\,[A/cm^2]$	Charge lifetime $Q_\tau[C]$	Fluence lifetime $F_\tau[C/cm^2]$
MAMI (achieved 2004 [1], pol)	$2 \cdot 10^{-4}$	0.2	200	$2 \cdot 10^5$
ERL based EIC (pol)	0.1	0.1	$3 \cdot 10^5$	$3 \cdot 10^5$
Electron cooler (unpol)	1	1	$3 \cdot 10^6$	$3 \cdot 10^6$
ERL light source (unpol)	0.1	10	$3 \cdot 10^5$	$3 \cdot 10^7$

One of the most prominent effects which tends to reduce τ is the so-called ion back-bombardment (IBB). IBB results from ionization of residual gas by the beam itself. In an electrostatic gun these ions travel backwards to the cathode, where they hit it with a kinetic energy that corresponds to the potential of the ionization site. In a homogeneous acceleration field only the emission site is affected by this impact. The rate of damage will be proportional to the ion current density on the surface which is itself $\propto$ to the electron current density: $j_{ion} \propto j_b = I_b/A$ (I_b= beam current, A=beam cross section), hence $\tau_{IBB} \propto j_b^{-1} = A/I_b$. Furthermore, the number of ions produced will be proportional to the residual gas pressure p_{res} and to the length of the acceleration stage which in turn is inversely proportional to the electric field (E). Putting all relations together we find : $\tau_{IBB} = kEA/(I_b p_{res})$ with a constant k that contains the sensitivity of the photocathode or its surface region against ion impact. The 'charge lifetime' is defined as $Q_\tau = I_b \tau_{IBB}$. We find:

$$Q_\tau = k\frac{EA}{p_{res}} \quad (1)$$
$$F_\tau = k\frac{E}{p_{res}}.$$

Q_τ and the 'fluence lifetime' $F_\tau = Q_\tau/A$ can be increased by reducing the gas pressure or by increasing E. A further conclusion from the relations is that the charge lifetime is supposed to be proportional to the emitting area. Nowadays operation at c.w. machines like CEBAF or MAMI is characterized by relatively low currents - fractions of milliamperes - which are drawn from small areas - $< 1\,mm^2$. The result in table 1 indicates that one seemingly approaches the requirements of an ERL based polarized EIC (EIC: Electron Ion Collider). The achieved F_τ is close to the requirements, which means that the desired charge lifetime is achievable if area scaling as stated in equation 1 would hold. This, however, needs experimental verification. One of the goals of our experiments was therefore to try to observe if there exists an increased charge lifetime for beams of larger beam diameter.

Comparing the achieved result with the requirements for an ERL based electron cooler or a light-source, we see that the existing performance is insufficient. However, these projects do not require polarized beam. This opens the possibility to improve the performance by using a NEA cathode with high energy photons. A probably more efficient approach is to resort to positive electron affinity (PEA) devices which should be much more stable, i.e. should show a larger value of k in eq. 1. Such a photocathode could be the well known KCsSb cathode which has an electron affinitiy of +1.1 eV and a photoemission threshold of 2.1 eV [2].

In the next section we present new apparative developments. We then describe the determination of other experimental parameters which also have influence on the observed lifetime. Section four presents the results achieved with NEA GaAs and an analysis of the q.e. pattern after the irradiation. We close with an outlook towards further experiments.

2. Apparative developments

2.1. Source with improved vacuum system

The 100 keV photosource used for our investigations is a back-up system which was buildt to replace the MAMI production source in case of a major failure. It has been optimized with respect to the specific pumping speed by designing a vacuum chamber which allowed to arrange 14 SAES WP700 modules (700l/s pumping speed for H_2 each) around the beam acceleration stage. This increased the specific pumping speed by about a factor 10. The HV parts remained unchanged to be compatible with the production source. The first 1.6 m long part of the beamline is electrically insulated, this region extends from a few centimeters behind the anode of the source towards about half a meter behind a 90 degree beam deflection by a "alpha" magnet. It is therefore possible to measure transmission losses in this part of the beamline with a resolution of nA, which can easily be checked if the beam is directed onto the beam pipe by miss-steering it. If the beam was properly adjusted no detectable loss was observed at 2 mA even for largest emission spot size on the cathode.

2.2. 405 nm Laser system

For experiments with unpolarized beam a 150 mW GaN Laser-diode was used which produced 405 nm light (3.06 eV photon-energy). We have tested this laser also in r.f. synchronized mode, which resulted in a 2.5 GHz pulse train with < 70 ps (FWHM) long pulses. Such a laser diode is an extremely simple means of injecting r.f. synchronized electron beams in c.w. accelerators with (unpolarized) average currents in the mA range. In spite of the small average power this is possible with NEA-GaAs since photosensitivities at 405 nm are $> 30mA/Watt$. The same holds for the potentially more stable $KCsSb$ cathodes where we have recently achieved $20mA/Watt$.

3. Corrections to observed lifetime

The observed lifetime τ in an experiment may be expressed as

$$\tau = \left(\frac{1}{\tau_{sum}} + \frac{1}{\tau_{IBB}}\right)^{-1}. \qquad (2)$$

The term τ_{sum} stands for sum of all other effects which tend to deteriorate the photocathode globally, i.e. on the whole active surface. In order to obtain a result for τ_{IBB} one has to correct for τ_{sum}. We therefore discuss the effects which determine τ_{sum}.

Vacuum lifetime and field emission: In figure 1 we present the q.e. as a function of time in 'armed and ready condition' when the 100 kV accelerating voltage is applied continuously and the valves in the beam lime are open. The resulting average current from periodic q.e. measurements is so small that it has no influence on the lifetime. The so measured decay constant characterizes the parallel interaction of field emission produced gases and ions and of the residual gas with the cathode surface. Starting with a period of almost constant q.e. an exponential decay sets in after about three weeks. Depending on weather one defines the lifetime as the time constant of the exponential fit or as the overall reduction to 1/e from the beginning we find $\tau_{sum} = 26 - 50$ days.

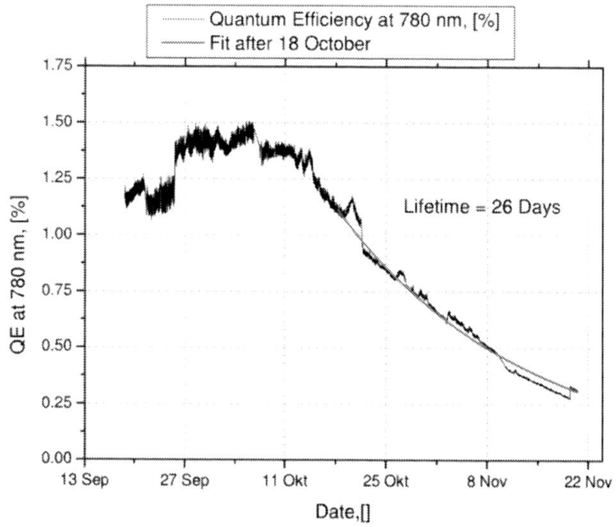

Figure 1. Quantum efficiency as a function of time with HV switched on and open valves in the beam line towards the beam dump but with negligible average current

Thermal effects: The value for τ_{sum} presented above is only an upper limit, since additional effects further diminish it in high current operation. For instance, our investigations of the dependence of τ_{sum} on cathode temperature [3] yield an exponential reduction:

$$\tau_{sum}(T) = \tau_{sum}(T_0 = 20°C)e^{(-((T-T_0)/50))}. \quad (3)$$

Because the cathode holders in our experiments may show thermal coefficients as low as $2mW/K$ (due to the low thermal conductivity of the interfaces between the semiconductor and its holder) we have to reduce laser intensities to below 100 mW to avoid excessive heating. This allowed for currents between 1 and 3 mA. The upper value is already limited by the capability of the 100 kV supply of the source. The measurements where performed with constant laser intensity to avoid a shortening of τ_{sum} because of increasing temperature.

Transmission losses: We have observed [4] that $\tau_{loss}[h] = 1000/I_{loss}[nA]$ gives an estimation for the effect of losses directly behind the anode. Hence, in our case, such losses must be reduced to about one nA - or a relative loss of $\approx 10^{-6}$. This requires to restrict the active surface of the cathode to a somewhat smaller circular diameter than the cathode itself - 3 mm vs. 10 mm - since emission from cathode regions close to its edge leads to losses which may reach $5 \cdot 10^{-4}$ if the whole cathode area is used. The restriction of the photo-active area is provided by the mask activation technique [5]. Under this condition we do not expect further reduction of τ_{sum} by beam losses. However, a factor that is difficult to quantify is the backstreaming of gas from the beam dump. In our set-up the beam dump is about three meters away from the source, it is separated from it by a 90 degree bend and backstreaming is suppressed by a three fold differential pumping stage. Pressure increase in the vicinity of the beam dump was $\approx 1 \cdot 10^{-9} mbar/mA$. If we assume a 10^3-fold reduction due to differential pumping, we get a 20% pressure-rise in the acceleration stage at 2 mA. This will reduce Q_τ directly via τ_{IBB} and will also reduce τ_{sum} due to the increased global interaction with residual gas.

Measurement of τ_{sum}: In order to avoid the uncertainties introduced by estimating the reduction of τ_{sum} due to increased temperature in operation and backstreaming we have tried to measure τ_{sum} by analyzing the q.e. distribution after operation. This resulted in a value of $\tau_{sum} = 15d$, which is in reasonable agreement with the expectation of a reduced τ_{sum} due to the increased temperature of the cathode during the experiment.

4. Results with NEA GaAs

Bulk GaAs cathodes were used at 780 nm (IR-light, 'polarized' operation) and 405 nm (blue light, 'unpolarized' operation). In polarized operation the source has roughly reproduced the values reported in table 1. With the blue laser we have performed two lifetime experiments at laser spot diameters of 0.3 and 1.5 mm respectively. The spot sizes are defined by the $1/e^2-$ values of the laser intensity distribution. The 1.5 mm result is presented in figure 2, both measurements yielded a nearly exponential decay, as indicated by the solid line in the figure which has the time constant $\tau = 5.5$ days with an initial current $I_0 = 2.3$ mA. The charge that is extracted during one lifetime under the condition of fixed laser intensity is $Q = I_0\tau(1 - 1/e)$, resulting in $Q(0.3mm) = 150\,C$ and $Q(1.5mm) = 700\,C$. The charge lifetime Q_τ (as defined in eq.1) is somewhat higher, since the value of τ_{IBB} instead of τ should be used. Extracting τ_{IBB} from equation 2 yields $Q_\tau(0.3\,mm) = 160 - 240C$[1] and $Q_\tau(1.5\,mm) = 1100C$. These results are in contradiction with area scaling since Q_τ is increased by a factor of only 4.5 – 6.9 whereas the ratio of emission areas is 25. This outcome is similar to an experiment which was published by

[1] The experiment at 0.3 mm was performed at lower laserpower and current (0.8 mA) which should yield a higher value of τ_{sum} if compared to the larger beam diameter. Unfortunately we have not been able to measure τ_{sum} in this case. We therefore assume: $15d < \tau_{sum} < 47d$, which leads to the variation in Q_τ after correction.

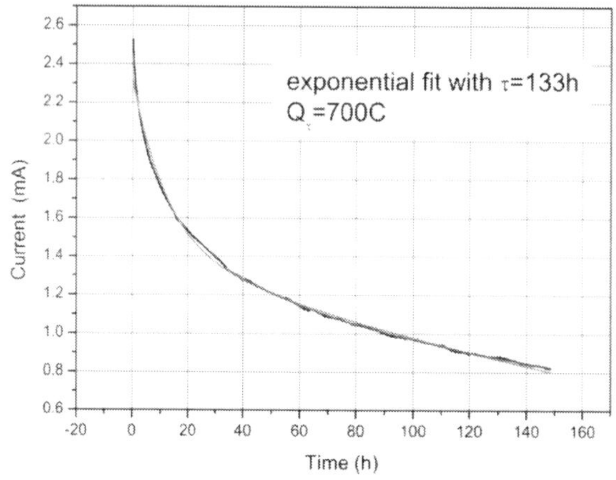

Figure 2. Current versus time at high intensity produced from 1.5 mm diameter laser spot.

Grames et al. from JLAB [6]. Larger absolute values of Q_τ - by about a factor of $\approx$ 2-3 - were observed. However, the discrepancy in the ratios of Q_τ and the emission areas was similar.

A part of the discrepancy may be explained by an analysis of the local q.e.-distribution after the experiment. Figures 3 and 4 present a map of q.e. before and after the production of 700 C of charge with 1.5 mm emission site diameter. A first interesting feature is the fringe of higher remaining q.e. in figure 4 where about 2/3 of the initial q.e. can still be found. Since these regions have only small overlap with the emission site, we may attribute the reduction in the fringe to global damage, i.e. τ_{sum}, resulting in $\tau_{sum} = 15d$.

Second, the region of maximum damage is not coinciding with the position of the laser spot(Cross in the right figure.). This feature is usually observed and is believed to be the result of a small displacement of the beam by about 1 mm from its starting position, maybe caused by magnetic stray fields in the acceleration stage or by the focusing effect of the conically shaped cathode holder. In contrast to the situation in a strictly homogeneous electrostatic field the electrons now follow a curved trajectory. The high energy ions which are produced towards the end of the acceleration are therefore projected back towards a position which is not exactly the same as the emitting spot. As a consequence there exists an area of about 1.5 mm diameter where the damage is very large. It seems that the size of this 'central damage area' is quite independent from the laser spot size. In regular accelerator operation beams of small diameter are started at a somewhat excentrical location in order to avoid the central damage area. Unfortunately, due to the smallness of the total active area - this restriction is necessary in order to avoid beam loss (section 3) - it is not possible to avoid an overlap of the emission site with this heavily damaged area in the case of the 1.5 mm beam, whereas it may well have been the case for 0.3 mm. This effect points in the right direction to remove some of the discrepancy. Furthermore, the high current density at small sizes - surpassing $2A/cm^2$ for the 0.3 mm experiment - may lead to a widening of the beam directly in front of the cathode. In consequence the increased effective beam area may also reduce the discrepancy.

5. Outlook

As a possible reason for the deviation from area-scaling we have identified the overlap of emission sites with $> 1mm$ diameters with the heavily damaged areas on the cathode. Since observations

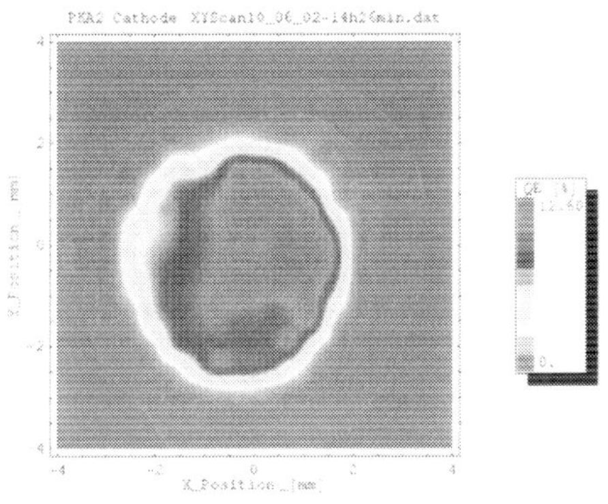

Figure 3. Quantum efficiency distribution before run

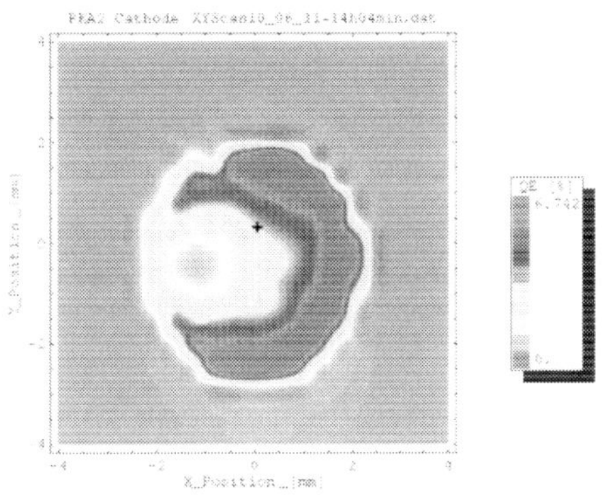

Figure 4. Same measurement after production of 700 C in the run.

so far support the assumption that the area of heavy damage is independent from the laser (emission-) size this effect can be mitigated by creating a somewhat larger activated area of $\approx$ 5 mm diameter. The larger size could allow to avoid the heavily damaged region for emission spot sizes of $\approx$ 2 mm diameter. It is desirable to produce the expected charges of about 5 kC within a time $<< \tau_{sum}$ in order to avoid uncertainties due to the corrections for τ_{sum}. Therefore it will be necessary to use currents of the order 20 mA. Before we can attempt such an experiment we have to solve two main problems: Most importantly the thermal conductivity of the mechanical interfaces in the cathode holder must be increased by about an order of magnitude. This is feasible by soldering the cathodes to the cathode holder and by pressing the holder with more force into the electrode structure. Second, the gas desorption in the beam dump must be further reduced. Probably the best way to achieve this is to replace the present faraday cup by the 'suppressed collector' of an electron cooler. Such devices achieve differential pressures of $< 10^{-11} mbar/mA$ which gives the one order of magnitude improvement necessary [7].

The more robust KCsSb photocathode will be tested in the near future. First cathodes with a sufficiently high quantum efficiency at 405 nm were successfully produced. They will be transported to the 100 kV source in UHV transport vessels. We expect to perform similar measurements as presented in this paper with KCsSb cathodes during 2011.

Acknowledgments

This work was supported by the Deutsche Forschungsgemeinschaft (DFG) within SFB 443 and by the Helmholtzgemeinschaft deutscher Forschungszentren (HGF) through the Helmholtz Institut Mainz (HIM).

[1] Aulenbacher K 2007 *Erzeugung intensiver hochpolarisierter Elektronenstrahlen mit hoher Symmetrie unter Helizitätswechsel* (Aachen: Shaker-Verlag) ISBN 978-3-8322-6429-1
[2] Sommer A H 1968 *Photoemissive Materials* (New York: Wiley and sons) ISBN 471-81300-1
[3] Riehn E, Tioukine V and Aulenbacher K 2009 *Proc. PST-2009* ed P Lenisa (Singapore: World scientific, (in press))
[4] Aulenbacher K, Euteneuer H, Jennewein P, Kaiser K H, Kreidel H J, DvHarrach, Reichert E, Schuler J, Tioukine V, Wiessner M and Winkler K (2001) *Spin 2000, 14th Int. Spin Physics Symp., AIP conf. proc. Vol. 570* ed Hatanaka K, Nakano T, Imai K and Ejiri H (AIP, Melville, New York) pp 949–954
[5] Aulenbacher K, Tioukine V, Wiessner M and Winkler K (2003) *SPIN 2002: 15th Int. Spin Physics Symp.; AIP conf. proc. Vol. 675* ed Makdisi K Y, AULuccio and WWMacKay (AIP, Melville, New York) pp 1088–1092
[6] Grames J, Poelker M, Adderley P, Brittain J, Clark J, Hansknecht J, Machie D, Stutzman M and Surles-Law K 2007 *Proc. Spin-2006, Kyoto* ed K Imai et al (Melville, New York: AIP, Vol. 915) p 1037
[7] H Poth 1990 *Physics Reports* **196** 135–297

Journal of Physics: Conference Series **298** (2011) 012020 doi:10.1088/1742-6596/298/1/012020

Polarized Positrons for Future Linear Colliders

Sabine Riemann, Andreas Schälicke, Andriy Ushakov
DESY, Platanenallee 6, D-15738 Zeuthen, Germany
E-mail: sabine.riemann@desy.de

Abstract. The high luminosity requirements and the option of a polarized beam present a great challenge for the design of the positron source for future linear colliders. This contribution provides an overview of the proposed designs for the polarized positron sources of the International Linear Collider (ILC) and the Compact Linear Collider (CLIC).

1. Introduction

Experiments at future linear electron-positron colliders running at high energies of few hundred GeV up to few TeV will allow to test the Standard Model and physics beyond. In contrast to hadron machines, the initial state parameters are known: the particle type is defined, and energy and beam polarization can be measured with high precision. Since weak interactions violate parity, beam polarization is important to differentiate weak processes and to search for new physics scenarios. The choice of beam polarization allows to enhance or suppress Standard Model processes and to disentangle new processes. Providing a high luminosity, precision measurements at high energies can be performed and will give us a deeper insight into the underlying symmetries and the nature of electroweak symmetry breaking. Although the physics potential can be substantially increased if both beams are polarized [1], up to now linear colliders have been operated with unpolarized positrons. It is a challenge to create the intense polarized positron beam for future e^+e^- machines as ILC and CLIC. However, already a low degree of positron polarization, $P_{e+} > 30\%$, improves the physics performance. In this contribution the possibilities to produce polarized positrons for future linear colliders are discussed and the main components of the positron source are presented.

2. Generation of polarized positrons

Longitudinally polarized positrons can be produced via the pair-production process initiated by circularly polarized photons [2]. Two methods are under consideration to produce the polarized photons for future linear colliders: the undulator scheme and laser backscattering. In the first scheme proposed by Balakin and Mikhailichenko [3] a multi-GeV electron beam is passed through a helical undulator to generate the multi-MeV circular polarized photon beam. In the second, circularly polarized laser light is backscattered off an electron beam [4, 5].

2.1. Polarized positrons from helical undulator

A helical undulator consists of a wire (often engineered as ribbon-wire) wound in a double helix to generate a rotating dipole field in the transverse planes along the beam axis. The electrons follow a helical path and emit photons. The photon spectra and photon radiation cone depend on the undulator parameters strength K and period λ, and on the electron energy. The desired photon yield determines the length of the undulator. Also the photon polarization spectrum depends on the undulator parameters; the direction of the helical field windings defines the sign of the polarization. The undulator method was tested in the proof-of-principle experiment E166 at the Final Focus Test Beam of the Stanford Linear Accelerator Center (SLAC).

2.1.1. Proof-of-principle experiment E166 The 46.6 GeV electron beam passed a 1-m-long helical undulator with period $\lambda = 2.25$ mm, $K = 0.17$ and aperture 0.9 mm. The circularly polarized photon beam of peak energy ≈ 8 MeV impinged on a 0.2-radiation-length tungsten target to produce positrons and electrons which were separated and measured using a Compton transmission polarimeter. In agreement with the expectations, the longitudinal polarization of positrons was above 80% for positrons at 6 MeV. A more detailed discussion of the experiment and its results can be found in references [6, 7].

2.2. The Compton scheme

Compton backscattering of circularly polarized photons off an electron beam results in longitudinally polarized electron-positron pairs. The sign of the averaged positron polarization is given by the the sign of the photon polarization. The method was tested in an experiment at the Accelerator Test Facility (ATF) at KEK [5].

2.2.1. Test experiment at KEK A circularly polarized laser beam of 532 nm was scattered off a high-quality, 1.28 GeV electron beam to produce photons with maximum energy of 56 MeV. A special apparatus (Compton chamber) was constructed to realize head-on electron-photon collisions. The magnitude of positron polarization was measured with a Compton transmission polarimeter for different states of laser polarization; $|P_{e+}| = 73 \pm 15(\text{stat}) \pm 19(\text{sys})\%$ was achieved. For linearly polarized laser light the positron polarization was zero. It is an advantage of Compton scheme that the reversal of positron helicity can be done easily by reversing the polarization of laser light.

2.3. Polarized positrons for future linear colliders

The methods described above allow to produce longitudinally polarized positrons from circularly polarized photons with an efficiency at the level of a few percent. Hence, a high power photon beam is needed to create the required amount of positrons. To keep the heat load at the positron production target within reasonable limits and to avoid damage, the target has to be rotated. Nevertheless, thermal problems in the target or the photon collimator demand special care in the design. Serious thermal problems also arise with a conventional unpolarized positron source where an electron beam hits a relatively thick target to produce electrons and positrons via the bremsstrahlungs process. It is possible to spread the heat load over a large area or several targets by deflecting the electron beam, but the thick target yields widely spread positrons resulting in a low capture efficiency. Therefore the energy deposition in the target and the capture section is very high, even higher than in photon-based positron sources.

3. The International Linear Collider (ILC)

The Reference Design Report (RDR) [8] for the ILC project was presented in 2007. The ILC energy should be adjustable from 200-500 GeV and upgradeable to 1 TeV, the luminosity amounts $2\times10^{34}\,\text{m}^{-2}\text{s}^{-1}$ to collect 500 fb^{-1} within the first four years. The electron beam will be polarized with a degree of at least 80%, positron polarization of about 60% is foreseen as an upgrade option. The energy stability and precision should be better than 0.1% to reach all physics goals. The layout is sketched in Fig. 1. The ILC includes the polarized electron source, the undulator-based

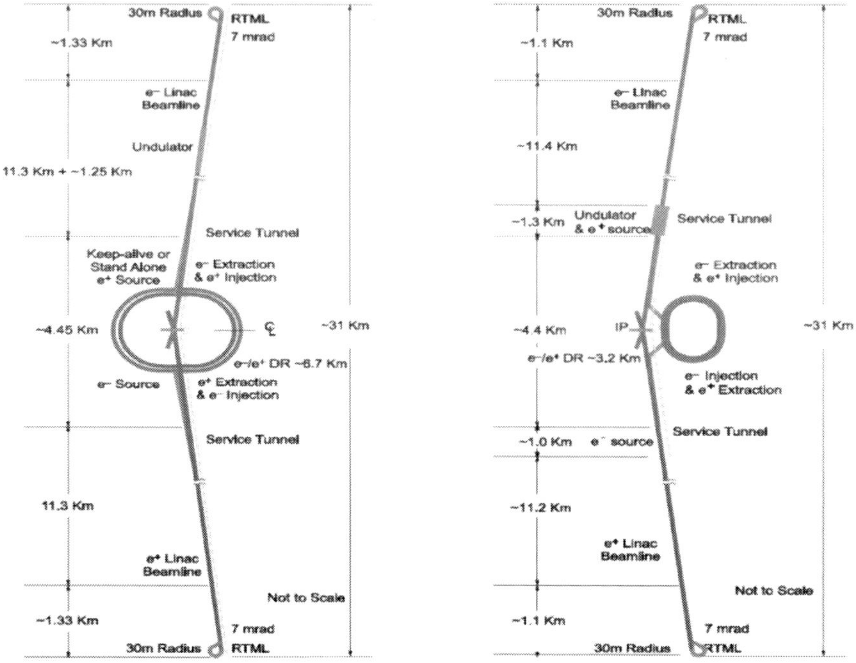

Figure 1. Schematic of the ILC (RDR left, SB2009 right).

positron source and the 5 GeV damping rings for electrons and positrons housed in a common tunnel at the center of the ILC. Each beam passes subsequently a bunch compressor system prior the injection into the 11 km long main linacs which are utilizing 1.3 GHz superconducting RF cavities operating at an average gradient of 31.5 MV/m. The RF pulse length is 1.6 ms, the pulse repetition rate is 5 Hz. The total power consumption is 230 MW. The ILC comprises one interaction region for measurements with two push-pull detectors.

The RDR design is revised for improvements and cost savings, currently the Strawman Baseline 2009 Design Proposal (SB2009) [9] is under consideration. The goal is to reduce the overall costs, to optimize the facilities and to improve the performance of the ILC. Main features are the change from the double-tunnel system to a single tunnel, the relocation of the positron source and a reduced power operation. The Technical Design Report will be released end 2012.

3.1. The ILC positron source

The positron source as described in the reference design [8] is based on a superconducting helical undulator located at the 150-GeV-point in electron linac. The undulator period is $\lambda = 1.15$ cm, the B-field on axis is 0.86 T corresponding to $K = 0.92$, and the aperture is 5.85 mm. The length of the undulator is 147 m to finally achieve the yield of 1.5 positrons per electron passing the undulator. The photons (10 MeV cut-off energy of the first harmonic) hit a 1.4 cm thick Ti-alloy target 500 m downstream the undulator. The positrons are captured using a pulsed flux concentrator (FC) as optical matching device and accelerated to 125 MeV using normal-conducting structures embedded in a solenoid with 0.5 T. The electrons and the photon beam are dumped, the positrons are accelerated to 5 GeV and injected into the damping ring. A 4 m ILC undulator prototype has been constructed and tested at Cockroft Institute [10]. The overall scheme of the ILC positron source is sketched in Fig. 2. This baseline design provides a

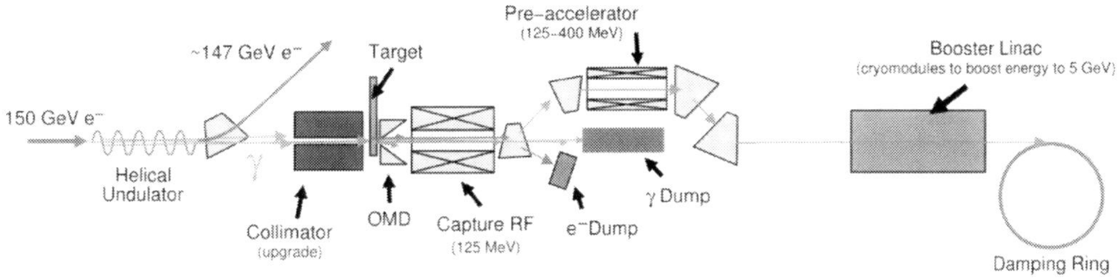

Figure 2. Schematic layout of the ILC positron source (not to scale).

positron beam with 34% polarization which can be further increased by implementing a photon collimator upstream the target [11]. To utilize the positron polarization for physics, it has to be brought to the interaction point, i.e. spin rotators are also needed for the positron beam line, and it must be possible to reverse the helicity of positrons rapidly to control systematic effects. Two proposals exist to change the helicity of positrons of energy 400 MeV or of 5 GeV [12]. A detailed spin tracking has still to be performed to finally decide the best option.

The SB2009 design suggests to place the positron source at the end of the main linac. This allows to use the same machine protection elements for both the undulator and the directly downstream beam delivery system, saving technical elements and tunnel space. However, this effects substantially the positron polarization: since the electron energy determines the opening angle of the undulator radiation, a 250 GeV-electron beam yields a positron polarization of only 22% (see also references [13, 14]). The physics gain of this low positron polarization is not sufficient taking into account the effort to keep it to the interaction point and to measure it precisely. So, the positron polarization should be increased, e.g. assembling a photon collimator or changing the undulator parameters. Alternatively, an unpolarized positron beam could be used; either by destroying the positron polarization or implementing a planar undulator to create linearly polarized photons. If they hit the target and undergo the pair-production process, unpolarized positrons and electrons are generated. But the photon yield in a helical undulator is about 1.5–2 times higher than that in a planar undulator for the K values of interest, not to mention that it is impossible to take advantage of polarized positrons for physics.

3.1.1. Positron target and capture The positron target is located about 500 m downstream the undulator and consists of a Titanium alloy wheel (2 m diameter) of thickness 1.4 cm and is designed as rim stabilized by five spokes. The incident photon beam spot size depends on the opening angle of the photon beam, it is of $\mathcal{O}(mm)$. The total power deposition in the target is 10.4 kW for the RDR and up to 8 kW for the SB2009 design. To reduce local thermal effects and also radiation damage, the target wheel has to be rotated with 2000 rpm.
The positron yield depends on the capture optics (optical matching device) behind the target. Assuming an Adiabatic Matching Device (AMD) with a B field tapered from 5 T at the target to 0.5 T at 20 cm downstream, the capture efficiency is 30%. However, eddy currents increase substantially the heat load in a rotating target immersed in a high magnetic field. The influence of eddy currents and the mechanical stability of the target wheel design were tested with a prototype (without cooling channels) built and operated at Cockroft Institute [15]. The torque associated with eddy current production in the target was measured depending on the immersion depth and the magnetic flux densities. The extrapolation of the results indicates that an additional power of 8 kW will be deposited in a target immersed in fields of approximately

1 T at revolution rates of 2000 rpm. Hence, the B field on the positron production target should be low to avoid eddy currents and the corresponding heat load. However, this implies a lower capture efficiency.
To minimize the risk, the SB2009 design suggests a quarter wave transformer (QWT) instead of the flux concentrator (FC) to collect the positrons exiting the target. This decreases the positron capture efficiency to 15%. A longer undulator (231 m) compensates this reduction. Using a QWT would make the positron polarization upgrade to 60% almost impossible. The undulator length has to be increased to an extent that the energy loss of electrons as well as the photon power loss in the undulator become worrying. A pulsed flux concentrator with low B field at the target affords a capture efficiency of 22% but requires still design work and the test of a prototype. It is under study at LLNL [16]. More details about the ILC positron source can be found in references [14, 17, 18].

4. The Compact Linear Collider (CLIC)

The Compact Linear Collider CLIC [19] is based on two-beam acceleration; the technology affords operation at 500 GeV, up to the multi-TeV range. The luminosity will be in the range $10^{34}\,cm^{-2}s^{-1}$. Energy and luminosity will be finally reviewed depending on the LHC physics results. The CLIC electron source will be polarized, polarized positrons are planned as an upgrade option.
The CLIC scheme uses high-frequency normal conducting accelerating structures with a main linac radiofrequency of 12 GHz and an accelerating field of 100 MV/m. The total power consumption should be below 500 MW.
The overall CLIC scheme is displayed in Fig. 3. The central injector complex includes sources,

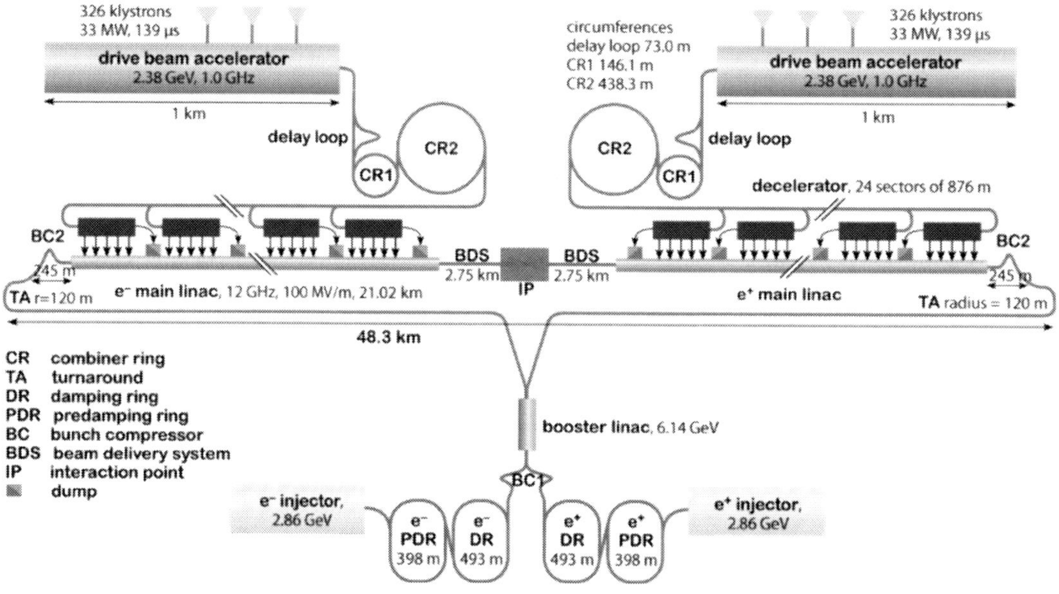

Figure 3. Schematic layout of CLIC 3 TeV (not to scale).

pre-damping and damping rings, bunch compressors and a 9 GeV booster. The injector complex prepares the ultra-low emittance beams which are needed to achieve the high luminosity. The radiofrequency power for the main linac is extracted from a high-intensity electron drive beam (100 A) with low energy (2.4 GeV). The drive beam is decelerated to 240 MeV, and the power is transferred by Power Extraction and Transfer Structures (PETS) to the low current (1.2 A) main beam for acceleration.

4.1. Polarized positrons for CLIC using the Compton scheme

The baseline positron source is unpolarized, polarization is planned as an upgrade option and needs substantial effort. The preferred design for the polarized positron source is based on the Laser-Compton method which can be implemented at any time without modifications of the CLIC complex. The problems concerning the positron production target and the capture section are similar to that studied for the ILC; the main differences concern the time structure and bunch charge. Hence, a close collaboration between the ILC and CLIC source communities exists.

Due to the relatively low Compton cross section high intensities for the electron and the laser beam are required which are presently not available. Several methods to produce the polarized positron beam with the CLIC time structure are under consideration, details can be found, e.g., in references [20, 21]:

- *CLIC positron injector with Compton ring* In this option a Compton ring with circulating electrons is used in conjunction with a high-finesse optical cavity powered by a YAG laser. Simulations show, that the size of the ring, the number of electrons per bunch and the length of the laser pulse can be synchronized such that sufficient 200-MeV-photons can be created to produce polarized positrons with the right pulse structure for CLIC.
- *CLIC Compton ERL* Instead of the Compton ring an Energy Recovering Linac can be used. This scheme provides a large number low-charge bunches with high-repetition frequency. The CLIC time structure is generated by a positron stacking procedure.
- *CLIC Compton linac* Here, the photons are created by Compton backscattering of photons from a CO_2 laser an electron beam from a linac. The CO_2 laser provides 10 times more photons per Joule than a YAG laser. The production of one photon per electron has been demonstrated at BNL [22]. To realize this scheme for CLIC, ten consecutive Compton IPs are needed to accumulate the required photon flux.

4.2. CLIC undulator scheme

The undulator option can also be applied at CLIC. But the integration of the undulator into the CLIC scheme has to be taken into account from the beginning, otherwise it is difficult to implement the undulator into the CLIC main linac and the positron injection part. The performance of the undulator, target and capture section are similar to that at the ILC, more details can be found in reference [21].

5. Polarimetry at the positron source

It requires some effort to measure the polarization at the positron source. Since the beam size is large upstream the damping ring, $\mathcal{O}$(cm) a Compton polarimeter will not work due to the tremendous laser power needed. It has been studied in simulations, that a Bhabha polarimeter would be a good option to measure the longitudinal polarization of positrons with energies of few hundred MeV [23]. Once the positrons passed a damping ring, a Compton polarimeter can be used. However, after passing the DR the spins are oriented to the transverse direction. The measurement of the asymmetries in the Compton scattering process with transversely polarized leptons depends on the energy and on the spatial distribution, and needs more effort than that of the Compton process with longitudinally polarized leptons. Since at CLIC the beams pass few damping rings in the injection complex, one should simulate the full spin tracking from the creation of positrons to the injection to the main linac to study possible sources of depolarization and to find out the best method and position for the polarimeter at the source.

6. Summary

Physics measurements at future high energy linear colliders will benefit from polarized beams. A positron source based on a helical undulator provides polarized positrons from the beginning, the degree of polarization can be increased by upgrades later on. Since later upgrades to a polarized source influence the source layout as well as the design of the whole accelerator, positron polarization must be regarded already in the baseline design.

The ILC design is already in a very progressive stage. Some problems are still under consideration; for the positron source these are in particular the thermal stress and the vacuum seals at the rotating positron production target as well as functionality of a flux concentrator to achieve a high positron yield. Currently the ILC layout is revised to develop the baseline design for the ILC TDR which will be released end 2012. The conceptual design of CLIC is under way and will be submitted in 2011. In parallel with the design work also a realistic spin tracking from start to end is being performed to figure out possible sources of depolarization and to guarantee a high and stable degree of positron polarization. Although not mentioned in this contribution, the precise measurement of the positron polarization at the interaction point is very important [24]. In particular, depolarization could occur in the strong field of beam-beam interaction during bunch crossing and has to be known with high accuracy. The success of a future LC depends substantially on making available polarized beams and the precise measurements at the IP.

Acknowledgment

We thank the organizers of PESP 2010 for the interesting and inspiring workshop. We acknowledge the ILC/CLIC positron source group team for fruitful collaboration.

References

[1] Moortgat-Pick G *et al.* 2008 *Phys. Rep.* **460**, 131-243
[2] Olsen H and Maximon L C 1959 *Phys. Rev.* **114**, 887-904
[3] Balakin V E and Mikhailichenko A A 1979 *Budker Institute of Nuclear Physics Report* No. BINP 79-85
[4] Bessonov E G and Mikhailichenko A A 1992 *Budker Institute of Nuclear Physics Report* No. BINP 92-43
[5] Omori T *et al.* 2006 *Phys. Rev. Lett.* **96** 114801
[6] Alexander G *et al.* 2009 *Nucl. Instrum. Meth. A* **610** 451-487
[7] Laihem K 2010 Contribution to this conference
[8] "International Linear Collider Reference Design Report: Volume 3: Accelerator" edited by Phinney N and Toge N and Walker N 2007 `http://www.linearcollider.org/rdr`
[9] `http://ilc-edmsdirect.desy.de/ilc-edmsdirect/file.jsp?edmsid=D00000000900425`
[10] Baynham E *et al.* 2009 *Proceedings of PAC09, Vancouver, BC, Canada* WE2RAI01
[11] Ushakov A *et al.* 2010 Contribution to this conference
[12] Moffeit K *et al.* 2008 ILC-NOTE-2008-040; Moffeit K *et al.* 2008 SLAC-TN-05-045
[13] Schälicke A *et al.* 2010 Contribution to this conference
[14] Clarke J A *et al. Proceedings of EPAC 08, Genoa, Italy, 2008* WEOBG03
[15] Bailey I *et al.* 2010 *Proceedings of IPAC'10, Kyoto, Japan* THPEC033
[16] Gronberg J *et al.* 2010 *Proceedings of IPAC'10, Kyoto, Japan* pp THPEC037
[17] Liu W and Gai W 2010 *Proceedings of IPAC'10, Kyoto, Japan* THPEC036
[18] Flöttmann K 2004 *Proceedings of EPAC 2004* TUZACH01
[19] Braun H *et al.* 2008 CERN-OPEN-2008-021 `http://clic-study.web.cern.ch/`
[20] Rinolfi L *et al.* 2009 *Proceedings of PAC09, Vancouver, BC, Canada* WE6RFP065
[21] Liu W *et al.* 2010 *Proceedings of IPAC'10, Kyoto, Japan* THPEC035
[22] Yakimenko V and Pogorelsky I 2006 *Phys. Rev. ST Accel. Beams* **9** 091001
[23] Alexander G *et al.* 2008 EUROTeV-Report-2008-091
[24] Moortgat-Pick G 2010 Contribution to this conference

Simulation of polarized positron sources for linear colliders

A. Ushakov, A. Schälicke, S. Riemann
DESY, Platanenallee 6, D-15738 Zeuthen, Germany
E-mail: andriy.ushakov@desy.de

Abstract. It is a challenge to provide an intense polarized positron source for future linear colliders. The positron yield, polarization and spin transport have to be optimized taking into account the energy deposition in the positron target and the activation of the source area. The Geant4-based simulation tool PPS-Sim has been developed for this purpose. Different production schemes, target options and matching devices can be configured and simulated. All source components and field parameters can be chosen easily. The main features of PPS-Sim as well as some selected simulation results are presented.

1. Introduction

The development of positron sources for the future electron-positron colliders requires detailed design studies. To perform simulations of positron beam generation and beam transport, a reliable software tool is needed that should include the variety of different source components.

For instance, for International Linear Collider (ILC) a helical undulator is planed as generator of circular polarized photons, that hit a titanium-alloy conversion target, and a pulsed flux concentrator as optical matching device will be used to collect the positrons. Additionally, an "auxiliary" conventional source will provide a positron beam with 1% intensity of the undulator-based source. For the CLIC positron source the so-called "hybrid-target" scheme is under discussion. First, the electron beam goes through a single crystal and generates photons via channeling effect. Then these photons are used to produce positrons in a second amorphous target. The Compton scheme is also considered as alternative option for the ILC and CLIC polarized positron sources.

Usually, the combination of two kind of tools are used to simulate the positron source. First, the interaction of the beam with the target is simulated (positron production). For this purpose Geant4 [1], EGS [2] or FLUKA [3] are used. The simulation of different positron source options (undulator-based, conventional, Compton source etc.) requires separate models for each kind of the source. In the second step the positrons are tracked through the electro-magnetic fields using codes as ASTRA [4], MAD-X [5], Bmad [6], BEAMPATH [7] etc.

The Geant4-based application "Polarized Positron Source Simulation" (PPS-Sim) is being developed to combine the simulation of different positron source options and the beam transport in a single software tool.

2. Description of PPS-Sim

PPS-Sim is based on the Geant4, ROOT and Qt4 libraries. Geant4 is used to simulate the electro-magnetic shower in the target, the polarization transfer, and the particle and spin tracking in electro-magnetic fields. ROOT is used to import external data, to store and analyse the simulation results. Qt4 libraries are used for development of the Graphical User Interface (GUI).

The positron source parts included in PPS-Sim are schematically shown in Fig. 1. The main source elements are the primary beam, the conversion target, the Optical Matching Device (OMD) and the accelerator cavity (RF).

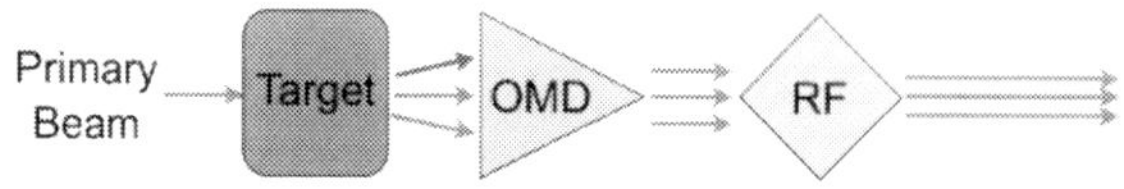

Figure 1. Schematic diagram of main source parts included in PPS-Sim.

The primary beam can be chosen between a photon beam generated in an undulator (undulator-based source), an electron beam (conventional source) or using an external ROOT-file representing the particle spectra of a Compton or hybrid-target source. PPS-Sim generates the undulator photons internally based on Kincaid's equations [10]. The necessary parameters for the generation of undulator photons are the energy of the drive electron beam, the undulator K-value, the undulator period λ, and the distance between undulator and target. The undulator-photon spectra (number photons generated by one electron passing one meter of undulator) and the polarization are shown in Fig. 2 for the ILC positron source with an undulator placed at 150 GeV (RDR design [8]) and the 250 GeV (SB2009 design [9]) electron beam energy.

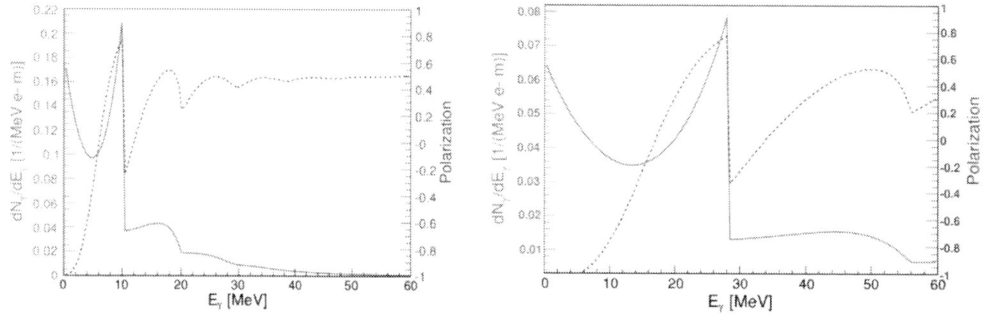

Figure 2. Energy spectra and polarization of photons generated in a helical undulator by 150 GeV electrons (left) and 250 GeV electrons (right). Undulator: $K = 0.92$, $\lambda = 11.5$ mm.

To enhance the polarization, a photon collimator can be placed between undulator and target. The collimator acts as a radial cut on the photon distribution and is not implemented as a real geometrical object.

Two kinds of conversion targets can be used: a solid wheel or a "liquid" target. The target material, thickness and other geometrical dimensions as well as the window parameters for the liquid target can be set by the user.

Three options of OMD have been implemented: adiabatic matching device (AMD), Lithium lens, and quarter-wave transformer (QWT).

The AMD is a tapered solenoid with high field strength at the beginning and low field strength at the end. The field along the beam (and solenoid) axis B_0 can be described as

$$B_0(z) = \frac{B_{ini}}{1 + gz},$$

where B_{ini} is the initial field and g is the taper parameter. The dependence of the positron yield defined as number of captured positrons that fulfil the damping ring (DR) requirements, and the dependence of the positron polarization on B_{ini} are shown in Fig. 3 for the ILC source (RDR design).

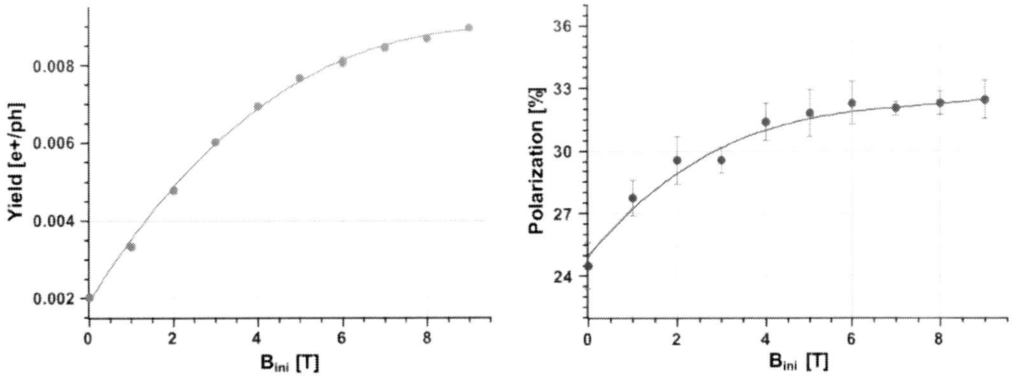

Figure 3. Dependence of yield (left) and polarization (right) on the initial field of the AMD (B_{ini}) for the ILC positron source (RDR design).

PPS-Sim does not include the whole beam line up to the DR. In order to estimate the number of positrons out of the DR acceptance, PPS-Sim applies cuts on the longitudinal bunch size and on the sum of x- and y-emittances (so-called diagonal emittance cut). By default, PPS-Sim applies a 10 mm longitudinal cut that corresponds to about 15° change of the electric field phase for 1.3 GHz.

The field of the Lithium lens implemented in PPS-Sim is described by an analytical function. Assuming that the current I in the lens has only a z-component and an equal density everywhere inside the lens, it is in cylindrical coordinates B_z, $B_r = 0$ and

$$B_\vartheta(r) = \frac{\mu_0 I r}{2\pi a^2}, \quad \text{for } r \leq a,$$

where μ_0 is the permeability of free space, and a the radius of the lens. A more realistic model of the lens is described in [11]. The focusing effect of the lens is shown in Fig. 4. In case of the ILC source (RDR undulator design), the optimal lens current, I_{LL}, is about 120 kA for a lens with 5 mm thickness and 8.5 mm radius.

The QWT consists of two solenoids. The first one has a higher and the second a lower field. Currently, PPS-Sim assumes that the field inside the solenoids is constant and in the region between them it is changed linearly. Figure 5 shows yield and polarization as function of the field strength of the first 2 cm long solenoid. The second 0.5 Tesla solenoid is placed 1 cm downstream the first one.

The E-field of the RF cavity embedded into a solenoid with constant B-field is modeled as harmonic function. The frequency, amplitude and phase of the E-field can be adjusted.

All parameters necessary to describe the fields, dimensions and the relative positions of the source parts can be configured in PPS-Sim via macro-files in batch mode or interactively in a GUI session. Figure 6 shows PPS-Sim running in interactive mode.

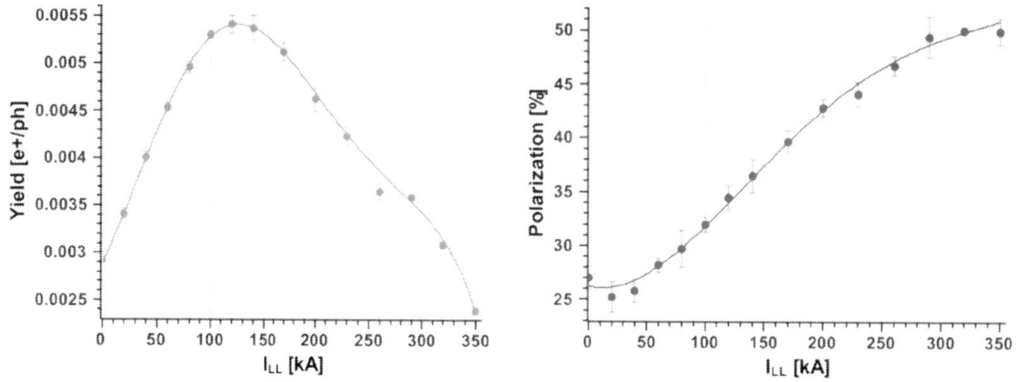

Figure 4. Dependence of yield (left) and polarization (right) on the Li-lens current (I_{LL}) for the ILC positron source (RDR design).

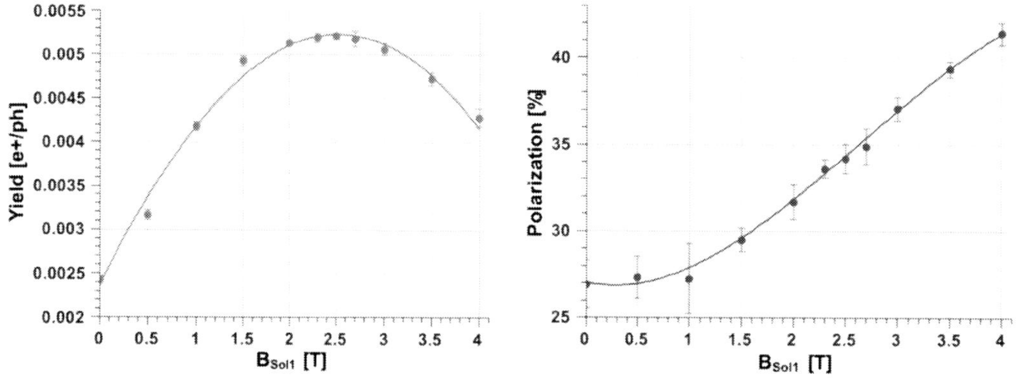

Figure 5. Dependence of yield (left) and polarization (right) on the field of the first solenoid (B_{Sol1}) of the QWT for the ILC positron source (RDR design).

Geant4 has a powerful geometry package and different possibilities to visualize the geometry of the model, particle trajectories and energy deposition in the positron source components. The 3D source model with liquid target and QWT is shown in Fig. 7. The user friendly visualization (OpenGL-based) of the geometry and particle trajectories is very useful for model development and debugging.

The ILC positron source should deliver $3 \cdot 10^{10}$ positrons in each of the 2625 bunches per pulse. To generate the required number of positrons, a photon beam with an average power of more than hundred kilowatt is required. The problem of energy deposition in a relatively small volume of the conversion target, development of shock-waves and radiation damage of the target have to be studied carefully in order to ensure a reliable lifetime of the target.

For the Peak Energy Deposition Density (PEDD) analysis, PPS-Sim provides a special running mode. The energy deposition inside the titanium target is shown in Fig. 8 for the ILC source with OMD yielding about 35% positron capture efficiency.

3. Summary

The development of Geant4 applications for positron source simulations allows to combine the positron production and tracking in one single tool. PPS-Sim is a flexible tool that includes a variety of source options. The undulator-based, conventional, Compton and hybrid-target

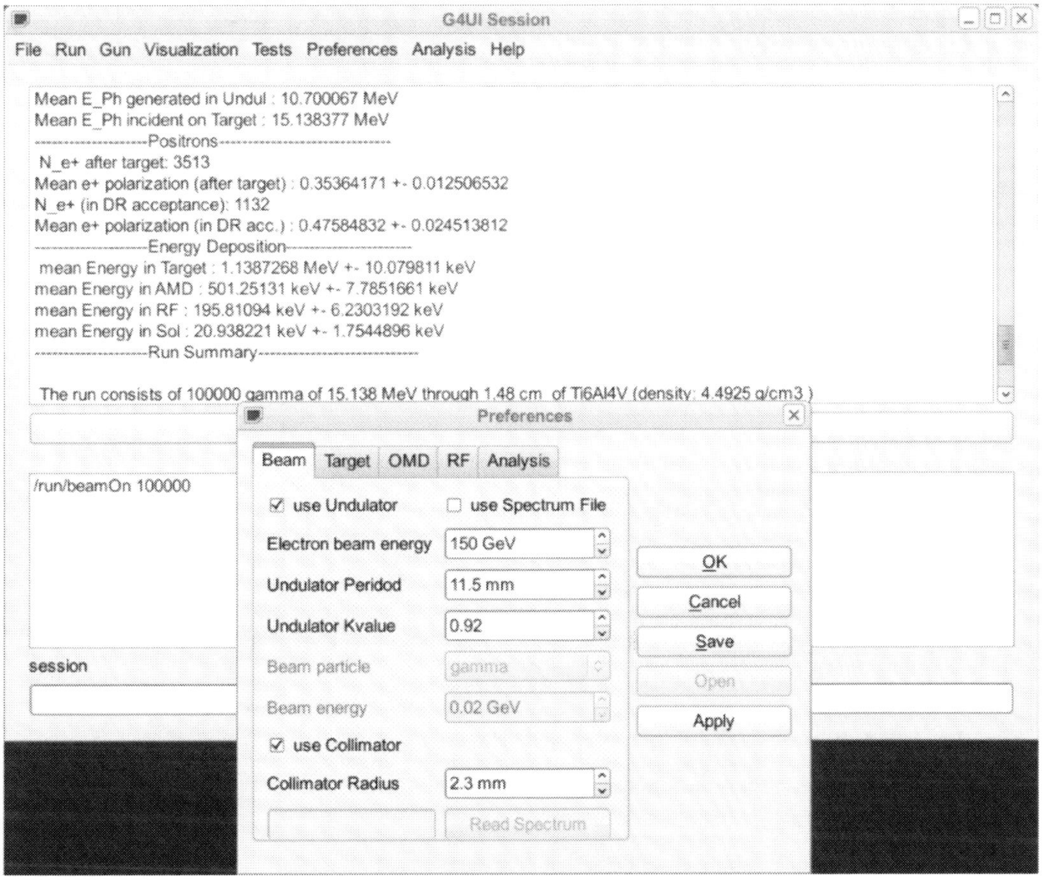

Figure 6. Examples of PPS-Sim windows: the window "Preferences" with opened tab for primary beam settings is placed over the main window "G4UI Session".

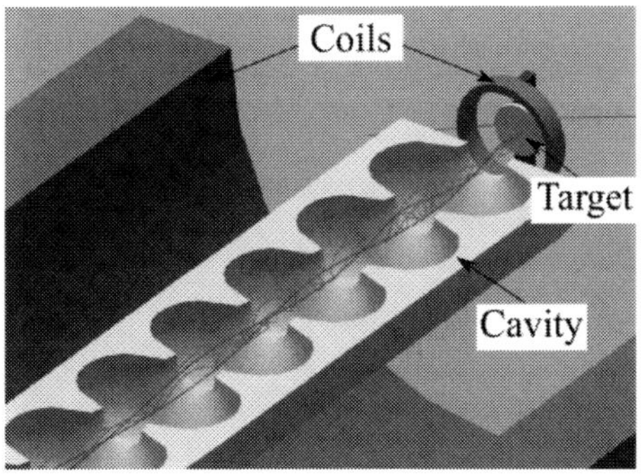

Figure 7. 3D view of the source model with liquid target (only liquid flow channel and windows are shown) and QWT (two coils). The electron trajectories are shown in red, positron trajectories in blue.

schemes of positron generation can be simulated and optimized. The different primary beams, conversion targets and OMD's can be selected. The graphical user interface is easy to use and to extend. The visualization of the geometry is useful for development and debugging. PPS-Sim

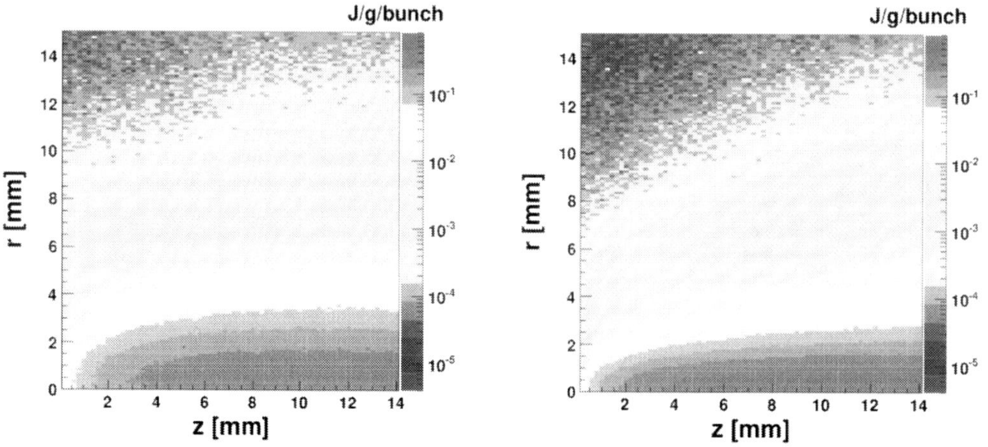

Figure 8. Spatial distribution of deposited energy in the titanium target of the ILC source: RDR design (left), SB2009 design (right).

is an open-source code and available for download from [12].

References

[1] Agostinelli S et al. 2003 *Nucl. Instr. and Meth.* A **506** 250
[2] http://www2.slac.stanford.edu/VVC/egs/
[3] Ferrari A, Sala P R, Fassò A and Ranft J 2005 CERN 2005-10
[4] http://wwww.desy.de/~mpyflo/
[5] http://mad.web.cern.ch/mad/
[6] http://www.lns.cornell.edu/~dcs/bmad
[7] Batygin Y K 2005 *Nucl. Instr. and Meth.* A **539** 455
[8] "International Linear Collider Reference Design Report: Volume 3: Accelelator" edited by Phinney N, Toge N and Walker N 2007
[9] http://ilc-edmsdirect.desy.de/ilc-edmsdirect/file.jsp?edmsid=D00000000900425
[10] Kincaid B M 1977 *J. Appl. Phys.* **48** 2684
[11] Mikhailichenko A 2010 *Cornell University Report* CBN 10-3
[12] http://pps-sim.desy.de/

Compton transmission polarimeter for a very precise polarization measurement within a wide range of electron currents

R Barday[1], K Aulenbacher[2], P Bangert[1], J Enders[1], A Göök[1], D H Jakubassa-Amundsen[3], F Nillius[2], A Surzhykov[4] and V A Yerokhin[5]

[1] Institut für Kernphysik, TU Darmstadt, Schlossgartenstr. 9, 64289 Darmstadt, Germany
[2] Institut für Kernphysik, Universität Mainz, J. J. Becherweg 45, 55118 Mainz, Germany
[3] Mathematisches Institut, Universität München, Theresienstr. 39, 80333 München, Germany
[4] Physikalisches Institut, Universität Heidelberg, Philosophenweg 12, 69120 Heidelberg, Germany
[5] Center for Advanced Studies, St. Petersburg State Polytechnical University, Polytekhnicheskaya 29, 195251 St. Petersburg, Russia

E-mail: barday@ikp.tu-darmstadt.de

Abstract. For new experiments with polarized electron and positron beams the precise and quick measurement of the beam polarization is required. The relative electron polarization has been monitored on-line using a compact transmission Compton polarimeter for a wide range of average electron currents up to $100\,\mu A$ and electron beam energy of $3.5\,\mathrm{MeV}$. The asymmetry induced by the beam polarization can be monitored precisely with $\Delta A/A < 0.2\,\%$. Even small relative variations of the beam polarization during a beam time may be detected using this polarimeter, such as an increase $\sim 1\,\%$ and subsequent decrease of the beam polarization.

1. Introduction

Different methods based on Mott and Møller scattering can be applied for polarization measurement. These polarimeters are destructive and cannot be operated during an experimental run. One has to assume that the polarization remains constant during the data taking or interrupt the experiment repeatedly for a polarization measurement. "Standard" Mott polarimeters are suitable up to electron beam energies of several MeV and are sensitive to the transverse beam polarization only. "Standard" Møller polarimeters are suitable at higher beam energies, but at low currents only (few μA), because of target heating. Another method is to analyse the circular polarization of the bremsstrahlung radiation produced when the electron beam hits a target. A simple way to measure the circular polarization of the photons is by determining their transmission through a magnetized absorber. Here one exploits the polarization dependence of the Compton scattering cross section: The transmitted intensity of circularly polarized photons scattered from the longitudinally polarized electrons in the magnetized absorber depends on the relative orientation of the photon and the electron spins. By switching the helicity of the photon beam, an asymmetry can be calculated.

2. Polarization transfer in bremsstrahlung

Circularly polarized bremsstrahlung radiation is produced when a polarized electron beam hits a radiator target. We will restrict ourselves to longitudinally polarized electrons, since for transversely polarized electrons the polarization transfer efficiency is smaller. This transfer efficiency depends on the energy E_γ and the emission angle θ of the photon. In the ultrarelativistic limit where the photons are nearly exclusively emitted in forward direction, Olsen and Maximon [1] have, based on approximate (Sommerfeld-Maue) electronic wavefunctions, derived the following expression for the circular polarization P_γ

$$P_\gamma = \frac{E_\gamma}{E_i} \frac{1 + \frac{1}{3}\left(1 - \frac{E_\gamma}{E_i}\right)}{1 - \frac{2}{3}\left(1 - \frac{E_\gamma}{E_i}\right) + \left(1 - \frac{E_\gamma}{E_i}\right)^2} P_e, \tag{1}$$

where $E_i = E_e + mc^2$ is the total energy of the electron beam and P_e is the polarization of the electron beam. Numerical calculations within the Elwert-Haug theory (for $E_\gamma < E_e$) performed at $E_e = 2 - 20\,\mathrm{MeV}$ and $\theta = 1^\circ$ show that equation (1) performs quite well (figure 1). In the short wavelength limit, $E_\gamma = E_e$, the polarization transfer is nearly complete at $E_e \geq 2\,\mathrm{MeV}$. From the Born approximation for bremsstrahlung, McVoy [2] has derived the following formula for circular polarization at the short wavelength limit and $\theta = 0^\circ$

$$P_\gamma^{\,max} = \left(1 + \frac{(1-\beta)\,(E_e + 2mc^2)}{(2-\beta)\,E_e}\right)^{-1} P_e, \tag{2}$$

which is plotted in figure 2 as a function of E_e. Also shown are numerical results from a more elaborate theory which uses Sommerfeld-Maue functions for the fast electron and Dirac functions for the slow one [3] at $E_e \geq 0.5$ MeV, respectively from partial wave calculations [4] at $E_e \leq 3$ MeV. Clearly, P_γ/P_e drops rapidly at beam energies below 1 MeV. We recall that neither of the formulae (1) or (2) depends on the nuclear charge Z of the radiator target. This Z-independence is, however, no longer true for low energies and when θ becomes large.

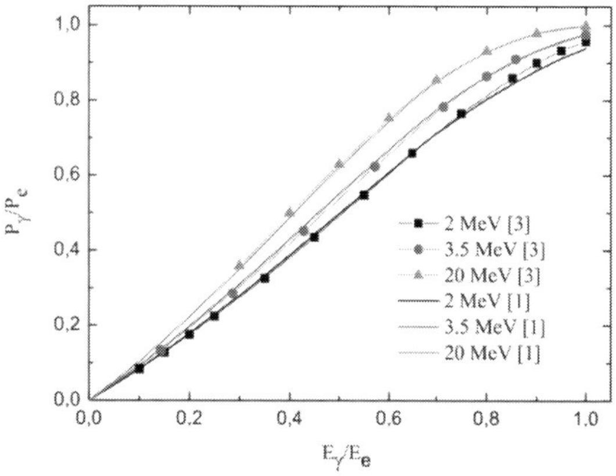

Figure 1. Circular polarization of bremsstrahlung radiation created by longitudinally polarized electron beam on Cu.

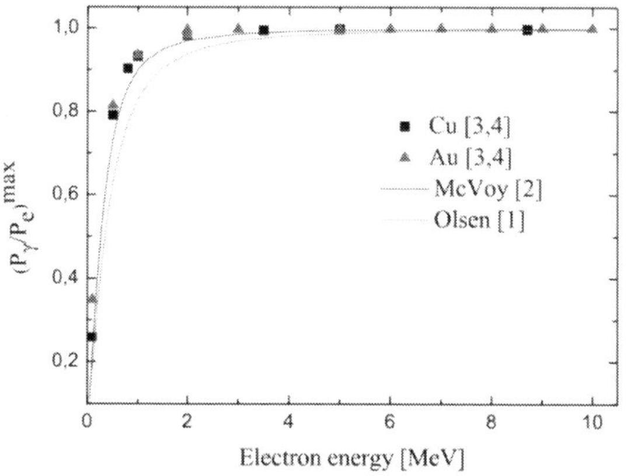

Figure 2. Efficiency of the polarization transfer as a function of the electron beam energy at the short wavelength limit.

3. Polarized Compton scattering

The Compton scattering of the photons with circular polarization P_γ from polarized electrons is given by [5]

$$\frac{d\sigma_c}{d\Omega} = \frac{r_0^2}{2}\left(\frac{k}{k_0}\right)^2 \{\phi_0 + fP_\gamma\phi_3\} \tag{3}$$

with $\phi_0 = 1 + \cos^2\theta + (k_0 - k)(1 - \cos\theta)$ and $\phi_3 = -(1 - \cos\theta)(k_0 + k)\cos\theta$. Here, r_0 is the classical electron radius, k_0 is the initial and k is the final photon momentum after scattering, ϕ_0 gives the ordinary Compton cross section, ϕ_3 is the polarization-dependent Compton cross section, θ is the scattering angle, and f is the fraction of longitudinally polarized electrons in the absorption magnet. By integrating the differential cross section, one obtains the total scattering cross section in the form

$$\sigma_c = \sigma_c^0 + fP_\gamma\sigma_c^{pol}, \tag{4}$$

where σ_c^0 is the polarization-insensitive (Klein-Nishina) and

$$\sigma_c^{pol} = \pm 2\pi r_0^2 \left\{\frac{1}{k_0} + \frac{k_0}{(1+2k_0)^2} - \frac{(1+k_0)}{2k_0^2}\ln(1+2k_0)\right\} \tag{5}$$

is the polarization-sensitive part. The term in the brackets changes sign at a photon energy of about 0.64 MeV and is negative at $E_\gamma > 0.64$ MeV. The photon transmission depends on the relative orientation of the photon and electron spins. The positive sign in equation (5) corresponds to the photon spin parallel to the electron spin and hence antiparallel to the magnetization. Or, in other words, for photon energies $E_\gamma > 0.64$ MeV the transmission is higher when the photon spin is parallel to the electron spin. The efficiency of the Compton transmission polarimeter depends on the ratio $|Z\sigma_c^{pol}/\sigma_{tot}|$, which has a maximum at 5 MeV (figure 3). Hence the measurement of the beam polarization using a Compton transmission polarimeter is most effective for energies of several MeV. At higher energies, for example at $E_e = 1.508$ GeV [6], the analyzing power of the Compton transmission polarimeter is very low, which makes an accurate polarization measurement challenging.

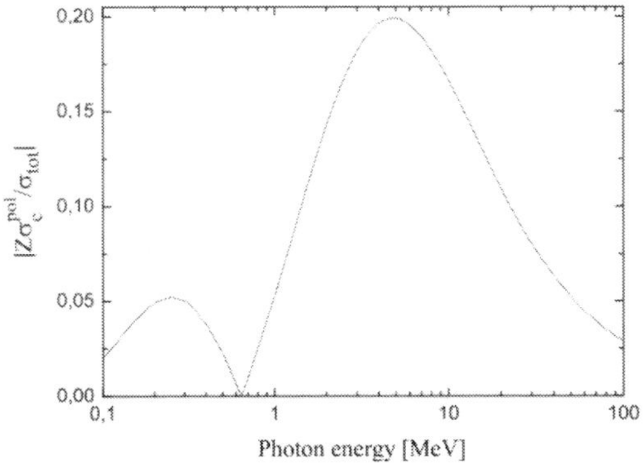

Figure 3. Ratio of the polarization dependent Compton cross section to the total cross section as a function of the photon energy for VACOFLUX50.

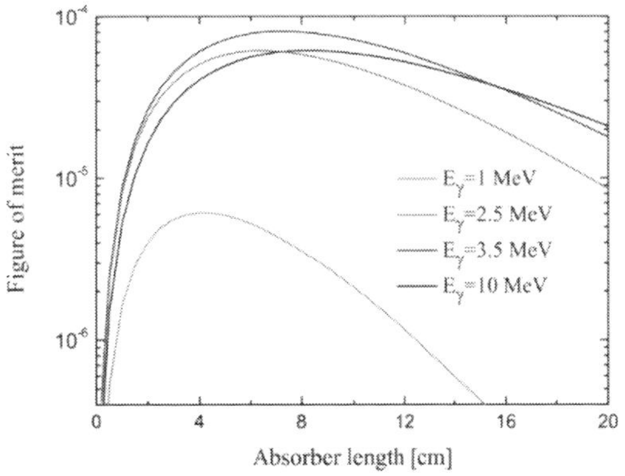

Figure 4. Dependence of the figure of merit on the length of the absorber for different photon energies.

The transmission probability for photons through the magnetized absorber of length L is given by

$$T^\pm = \exp\{-nL\sigma_{tot}\} = \exp\{-nL(\sigma_{photo} + \sigma_{pair} + Z\sigma_c)\}, \tag{6}$$

where n denotes the number density of atoms in the absorber. The asymmetry is then given by

$$A = \frac{T^+ - T^-}{T^+ + T^-} = \tanh\left(-nLZfP_\gamma\sigma_c^{pol}\right) \sim \left(-nLZfP_\gamma\sigma_c^{pol}\right), \tag{7}$$

because $nLZfP_\gamma\sigma_c^{pol} < 0.1$. The asymmetry in transmission increases linearly with increasing length of the absorber, but the intensity passing the absorber falls exponentially. If the counting rate is the limiting factor for the accuracy of the measurement, the optimum length $L_{opt} = 2\,(n\sigma_{total})^{-1}$ may be found from minimizing $\Delta A/A$, where ΔA is the statistical error in the asymmetry A and σ_{total} is the total absorption cross section. A figure of merit for the polarimeter A^2T for different photon energies as a function of the length of the absorption magnet is plotted in figure 4. Calculations are done using a photon polarization of 80 % and an electron polarization in the absorber of $f = 8$ %. If the counting rate is large, the length of the absorber can be increased to enhance the asymmetry. The absorber of the polarimeter described below consists of a 9 cm core and return yokes. The asymmetry of the transmitted intensity calculated using equation (7) for a 9 cm absorption magnet and the results from figure 1 for P_γ are shown in figure 5 as a function of the photon energy.

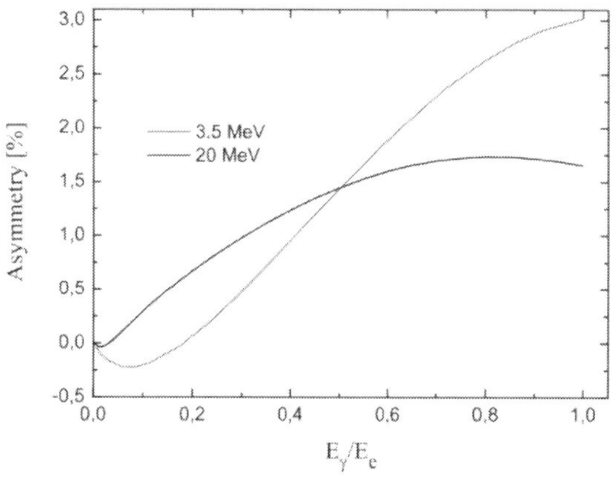

Figure 5. Predicted asymmetry as a function of the photon energy for the incident electron beam energies of 3.5 and 20 MeV.

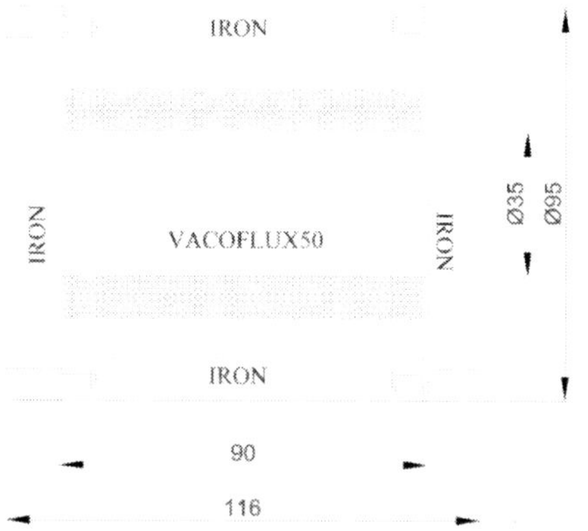

Figure 6. The absorption magnet with 420 turns, symbolized by the rows of circles.

4. Polarimeter

The Compton transmission polarimeter consists of a radiator, which produces the bremsstrahlung radiation, two CsI(Tl) detectors to measure the intensity of the photons and a magnetized absorber. As a radiator target we have chosen a copper target with a length of 12 mm. Our simulation shows that not more than 0.04 % of the incident electrons penetrate through the radiator. The electron polarization in the magnetized absorber is determined by

$$f = \frac{M_s}{N_e\mu_B} \sim \frac{2M}{N_e\mu_B}\left(\frac{g'-1}{g'}\right), \tag{8}$$

where M_s is the magnetization due to electron spin, M is the total magnetization, N_e is the electron density, μ_B is the Bohr magneton, and g' is the magnetomechanical ratio. The core of the absorption magnet is made from VACOFLUX50 (49%Co, 1.9%V and 49.1%Fe). VACOFLUX50 has a Curie temperature of 1223 K and an electron polarization of 8.5% at

2.35 T. High magnetization is achieved through annealing in a dry hydrogen atmosphere at 820^0C. The dimensions and shape of the absorption magnet are shown in figure 6. The ends of the core are machined with high precession to be flat and perpendicular to the axis.
Because of the photomultipliers saturation at high electron currents and high operation voltage, we use a silicon PIN photodiode (SPD) to read out the scintillation light. A combination of a scintillator with a SPD has a number of advantages for the photon detection, like high detection efficiency, high light output, temperature dependence of the signal only in the crystal part, insensitivity to magnetic fields and compactness. To read out the scintillation light of CsI(Tl), with a peak emission wavelength at 550 nm a Hamamatsu S3204 PIN-photodiode with an active area of $18\times18\,\mathrm{mm}^2$ is used. The quantum efficiency of the SPD at 550 nm is about 80 %. The CsI(Tl) scintillator was purchased from Saint-Gobain and has dimensions of $17\times17\,\mathrm{mm}^2$ and 50 mm length. After polishing all faces of the scintillator, it was wrapped in Teflon, followed by aluminium foil. The optical connection between the photodiode and the crystal is a RTV based silicon glue manufactured by GE Bayer Silicones. The energy resolution of the detector is about 6 % FWHM for 1332 keV.

5. Polarization measurements

The experiment has been performed at the MAMI injector [7]. A polarized electron beam is produced by illuminating a $GaAs/GaAs_{0.64}P_{0.36}$ strained superlattice photocathode [8] with circularly polarized light of 776 nm. The electron beam is accelerated to 3.5 MeV and deflected to 30^0 by the dipole magnet for the polarization measurement, as is shown in figure 7.

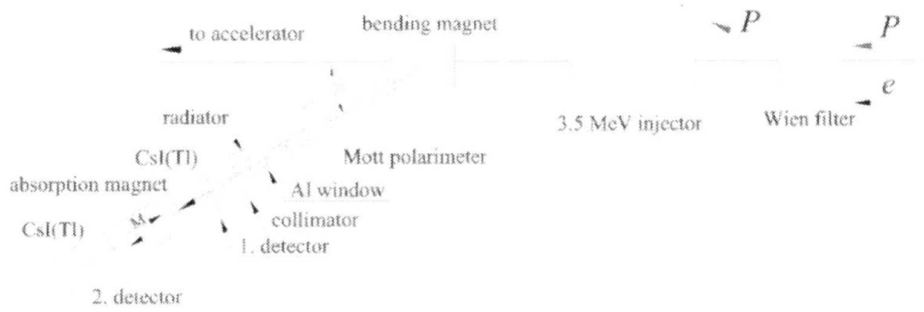

Figure 7. Layout of the experimental area.

The asymmetry of the transmitted photons determined in the experiment is given by

$$A = \frac{I_2^+/I_1^+ - I_2^-/I_1^-}{I_2^+/I_1^+ + I_2^-/I_1^-}, \tag{9}$$

where $I_1^\pm$ and $I_2^\pm$ are the signals from the first and second detector for two different helicities of the beam, respectively. We measured the dependence of the asymmetry on the current $I_{absorber}$ in the absorption magnet (figure 8). The electron beam current was $8\,\mu A$, and the electron spins were oriented at 45^0 with respect to the beam direction using the Wien filter to control the beam polarization using a Mott polarimeter additionally. Each measurement took about 60 seconds. One recognizes the hysteresis effects. About 1100 Ampere turns are needed to saturate the absorber. To limit heating of the absorption magnet it was magnetized in the later experiments using a current of 2.8 A.

The dependence of the asymmetry on the average electron beam current was also investigated. The polarization vector was, again, oriented at 45^0 with respect to the beam orientation. The transmission between the gun and the polarimeter in this experiment was approximately 90 %. We observed a slight change of the asymmetry of about $\Delta A/A \sim 10^{-4}$ (figure 9(a)) per μA

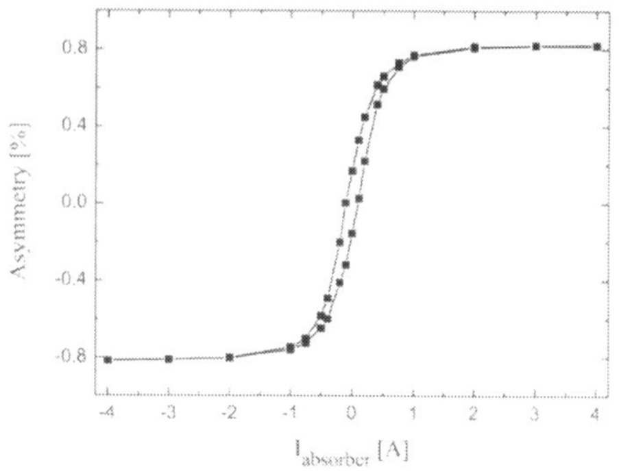

Figure 8. "Hysteresis loop" measured with polarized electron beam.

of beam current. This could be an effect of the thermal heating of the photocathode. The cathode holder is not optimized with respect to the heat transfer. Most of the laser intensity will be transferred to the crystal lattice of the photocathode, causing a temperature increase. This might affect the beam polarization. To see how temperature affects the spin polarization, consider the polarization dependence on the relaxation time

$$P = \frac{P_0}{1 + \tau/\tau_s} B_{BBR}, \tag{10}$$

where P_0 is the initial electron polarization in the conduction band, τ is the photoemission time of electrons, τ_s is the spin relaxation time, and B_{BBR} represents the depolarization in the band-bending region. The photoemission time τ amounts to several picoseconds and is approximately temperature independent, but τ_s decreases with increasing temperature [9]. Hence it can be expected that the polarization decreases at higher currents.

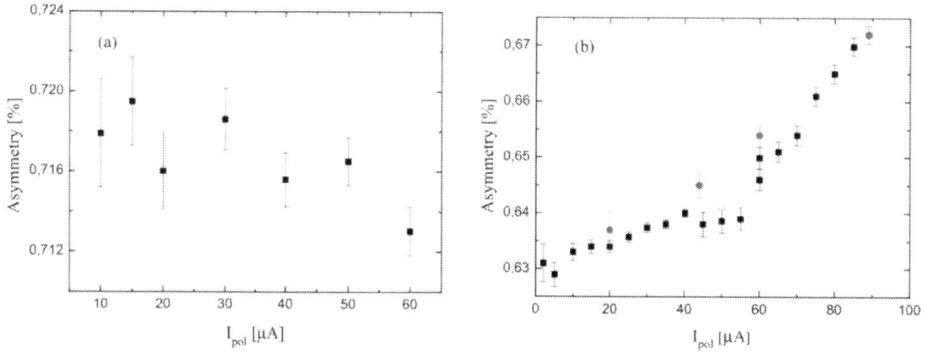

Figure 9. Observations of Compton asymmetry vs beam current I_{pol} at polarimeter. Black datapoints in left and right figures are subsequent measurements with an interruption of three months in between them. I_{pol} variation for the black points was obtained by varying the laser intensity on the cathode, the 'red' dataset in the right figure was obtained by reducing the beam current, produced by a constant laser intensity, by a variable electron beam collimator slit.

Three months later we repeated the experiment. The first detector was irradiated with gamma rays with about 10 kGy. The change of the asymmetry for currents below $60\,\mu A$ is of the same order as three months ago, but has an opposite sign, perhaps because of the radiation damage.

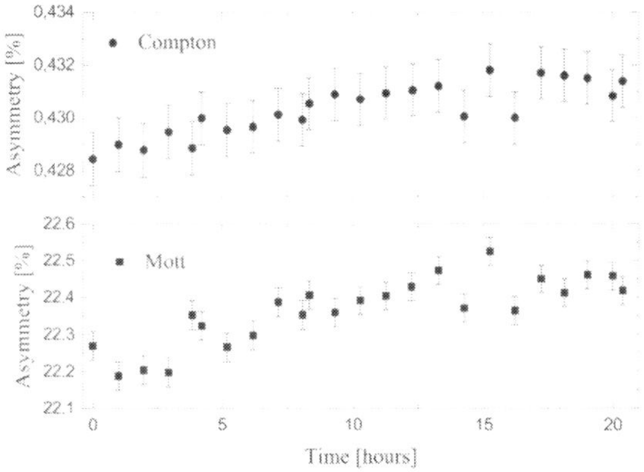

Figure 10. Variation of the electron spin polarization during a beam time, measured with Compton and Mott polarimeter.

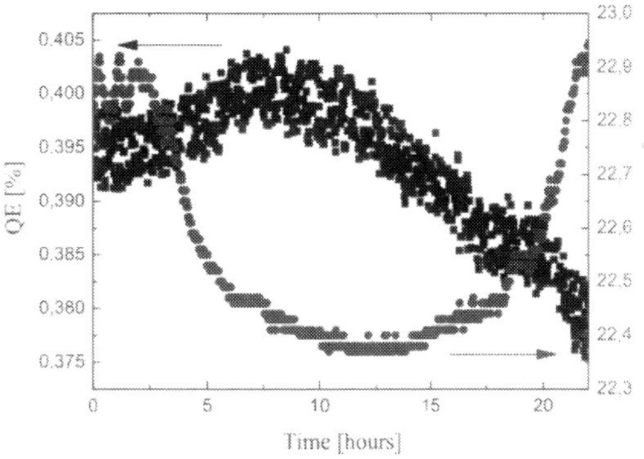

Figure 11. Quantum efficiency (squares) and temperature (circles) near the gun.

At higher beam currents the change of the asymmetry is significantly greater. The non-linearity of the photon detectors is responsible for this effect. To show that the change of the asymmetry in figure 9(b) is not an effect of the photocathode, we increased the laser power to produce $100\,\mu A$ average current from the gun and kept it constant. The beam current $I_{\rm pol}$ delivered to the polarimeter was reduced with a movable collimator slit in the chopper system between the source and the first accelerating structure. For both types of variation the current dependence is very similar, which provides strong evidence that the variation is not caused by intensity dependent effects - such as cathode heating - in the photocathode.

The long-term variation of the electron spin polarization was studied, as well. Data taking started 23 hours after cathode activation. The electron beam current delivered to the polarimeter was $30\,\mu$A, the polarization angle was 45^0. In order to make the Compton and Mott measurements simultaneously, we moved a 250-nm-thick gold target into the Mott polarimeter. Both polarimeters show an approximately linear increase of the polarization 15 hours after the start of the measurement (figure 10). During this experiment the quantum efficiency of the photocathode and the temperature near the gun were monitored (figure 11). The maximum polarization was observed 7 hours after the maximum of the quantum efficiency at a QE of $\sim 97\%$ of the maximum value. A similar behaviour of the polarization was observed in [10].

Acknowledgments
The authors want to thank C. Weinrich for helpful discussion. This work was supported by the Deutsche Forschungsgemeinschaft through SFB 443 and SFB 634 and by the state of Hesse through the LOEWE centre HIC for FAIR.

References

[1] Olsen H and Maximon L C 1959 *Phys. Rev.* **114** 887
[2] McVoy K W 1957 *Phys. Rev.* **106** 828 (Erratum: 1958 *Phys. Rev.* **110** 1484)
[3] Jakubassa-Amundsen D H 2010 *Phys. Rev.* A **82** 042714
[4] Yerokhin V A and Surzhykov A, submitted to publication
[5] Schopper H 1958 *Nucl. Instr.* **3** 158
[6] Weinrich C private communication and 2005 *Eur. Phys. J.* A **24** s2 129
[7] Aulenbacher K 1997 *Nucl. Instr. and Meth.* A **391** 498
[8] Maruyama T 2005 Proc. 16th Int. Spin Symp. SPIN2004 (Singapore: World Scientific) 917
[9] Kurebayashi H 2010 *Appl. Phys. Lett.* **96** 022505 and Zerrouati K 1988 *Phys. Rev.* B **37** 1334
[10] Mamaev Yu A 2001 AIP Conf. Proc. vol 570 p 920 and Tawada M 1997 *Jpn. J. Appl. Phys.* **36** 2863.

Precise measurement of the longitudinal polarisation at HERA with a Fabry-Perot cavity polarimeter

Marie Jacquet
Laboratoire de l'Accélérateur Linéaire, Orsay, France
E-mail: `mjacquet@lal.in2p3.fr`

Abstract. A Fabry-Perot cavity polarimeter installed in 2003 at HERA was designed to measure the longitudinal polarisation of the electron beam with high precision for each electron bunch spaced with a time interval of 96 ns. By interacting a high intensity laser (enhanced up to a few kW within the cavity) with the HERA electron beam it is possible to measure its polarisation with a relative statistical precision of 2% per bunch per minute. Detailed systematics studies have been performed and in particular a complete theoretical model has been developed in order to control at the per mill level the degree of circular polarisation of the laser beam. The result is a total systematic uncertainty of 1%. This is the first time that such a precision is achieved in the difficult, hostile and noisy environment of a particle collider.

1. Introduction

A fast and high precision longitudinal Compton polarimeter using a continuous wave laser resonating in a Fabry-Perot cavity (LPOL cavity) was proposed, constructed and installed in the electron-proton collider HERA near the existing longitudinal Compton polarimeter (LPOL) [1]. In addition to the LPOL, the transverse polarisation of the electron beam is also measured by another polarimeter (TPOL) [2].

With respect to the prior HERA LPOL and TPOL polarimeters, the higher statistical precision of the LPOL cavity is achieved by increasing firstly the power of the continuous wave laser by more than two orders of magnitude compared to the TPOL and secondly the frequency of the electron-photon(laser) interaction to 10 MHz compared to 0.1 kHZ of the pulsed laser of the LPOL. A new Data Acquisition System (DAQ), synchronised to the HERA beam clock, has been developed accordingly which operates without any trigger at 10 MHz.

Among all the source of systematic errors, the knowledge of the degree of circular polarisation S_3 of the laser beam at the electron-laser interaction point (IP) is of particular interest. Indeed, this quantity is directly involved in the following Compton scattering cross section:

$$d\sigma/dE_\gamma \;=\; d\sigma_0/dE_\gamma \;-\; P_z S_3 \; d\sigma_L/dE_\gamma \qquad (1)$$

(where E_γ is the energy of the scattered photon and σ_0 and σ_L are respectively the non-polarised and the longitudinal Compton cross sections), and as shown in Eq.(1) only the product S_3P_z is determined in Compton polarimetry. S_3 has therefore to be measured precisely in order to achieve the same level of precision for P_z.

The main purpose of the present article is to report about the electron polarisation measurement with the LPOL cavity experiment, and describe the experimental setup and methods that we have used to reach a few per mill level of systematic uncertainty on S_3.

2. LPOL cavity principle and electron polarisation extraction

The HERA Fabry-Perot cavity is similar to a device that has been used successfully to measure the polarisation of the CEBAF LINAC electron beam [3, 4, 5]. One major difference between the HERA and CEBAF LPOL cavities is the dynamical regime. Whereas the luminosity of Compton scattering is relatively low at CEBAF, it reaches much higher values at HERA. That is, the average number of back-scattered Compton photons (hereafter named BCP) is close to one per bunch in the latter case and much smaller in the former. The HERA dynamical regime, denoted as 'few photon mode' in this article, has been used successfully for the first time by the LPOL cavity to measure the electron beam polarisation. The advantage of the few photon mode is that one can calibrate the calorimeter in an absolute way using two reference points of the photon energy spectrum independently of the electron beam polarisation: firstly the Compton kinematic edge, located at 10 GeV at HERA for our 1 064 nm laser beam wavelength, and secondly the bremsstrahlung kinematic edge, located around 27.5 GeV which corresponds to photons radiated from the scattering of the electron beam with the residual gas of the vacuum beam pipe (hereafter named BGP).

The LPOL cavity overall system is shown in figure 1. Each of the 220 electron bunches crosses the laser beam and the BCPs reach a calorimeter at about 60 m from the electron-laser IP. A precise description of the cavity, the calorimeter, electronics and DAQ can be found in [6, 7]. The available data for the polarisation measurement of an individual bunch consists

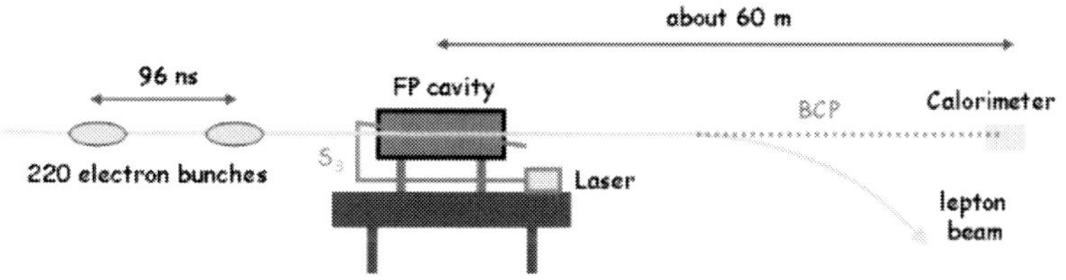

Figure 1. Overview of the area of the LPOL cavity installed in the HERA tunnel.

of a pair of photon energy histograms. These are successively recorded by the calorimeter DAQ for each bunch during one DAQ period (≈ 10 s) for the two polarisation states of the laser beam $S_3 = +1$ and $S_3 = -1$. These spectra come from a sum of genuine BCPs from electron-laser interactions, and three main backgrounds: the already mentioned BGP, the electron beam scattering off Black Body Photons (hereafter named BBP) emitted by the hot beam pipe, and the Synchrotron Radiation Photons (hereafter named SRP). Since the DAQ operates without trigger, the total number of entries of the histograms are fixed to 400 000 (corresponding to the number of HERA turns accumulated during one DAQ period).

To extract the electron polarisation from a pair of bunch energy histograms, the theoretical energy spectra of one BCP (see Eq.(1)), BGP [8, 9] and BBP [10] are computed numerically. The SRP background [11] is considered as a pedestal and it is treated separately. All these spectra are mixed and the convolution of all the processes is performed to provide an energy histogram. The detector effects (calorimeter, ADC/energy conversion and electronic noise) are applied to the energy histograms. A comparison between the experimental and calculated energy histograms is performed using a likelihood fit. The unknown parameters of this fit are the luminosities of the BCP, BGP and BBP processes integrated over one DAQ period, the electron beam polarisation and the parameters describing the detector effects. A detailed derivation of the formula used in our fits can be found in [7, 12]. In figure 2, such a calculated energy histogram is shown together with the experimental one. An excellent agreement is seen between the data and the red curve representing the sum of all processes. In figure 3, the two spectra for laser polarisation $S_3 = 1$ and $S_3 = -1$ are shown in the 'Compton energy range' together with the measured values. The difference allows the electron polarisation measurement.

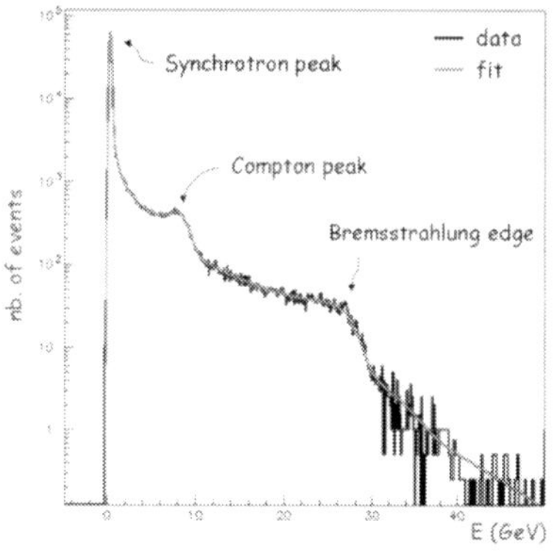

Figure 2. Measured energy spectrum (black) accumulated during 10 s of acquisition time and fitted result (red).

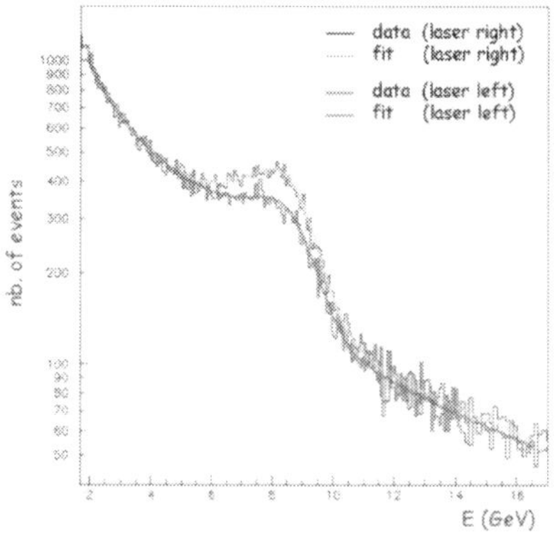

Figure 3. Compton edges of two energy spectra where the laser helicity was $S_3 = 1$ (right) and $S_3 = -1$ (left).

The total LPOL cavity data taking amounts to about 500 hours (from 6 October 2006 to the end of the HERA running in June 2007). The LPOL cavity provides a bunch dependent polarisation measurement every 20 s. The relative statistical precision is about 2% per bunch per minute as shown in figure 4 for a typical example. All LPOL cavity data have been analysed with the polarisation results available to be compared with the corresponding measurements from the TPOL. Figure 5 shows an example of polarisation measurements of one HERA luminosity fill provided by the LPOL cavity and TPOL polarimeters for the colliding electron bunches. The better statistical precision from the LPOL cavity polarimeter can be clearly appreciated.

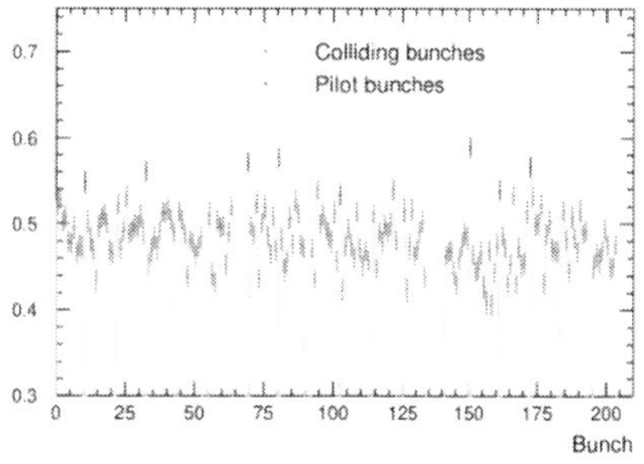

Figure 4. Bunch dependent polarisation measurement from the LPOL cavity averaged over 3 independent measurements corresponding to about one minute duration. The error bars represent the statistical precision of the measurement with the solid (dashed) histograms showing the colliding (pilot) bunch structure.

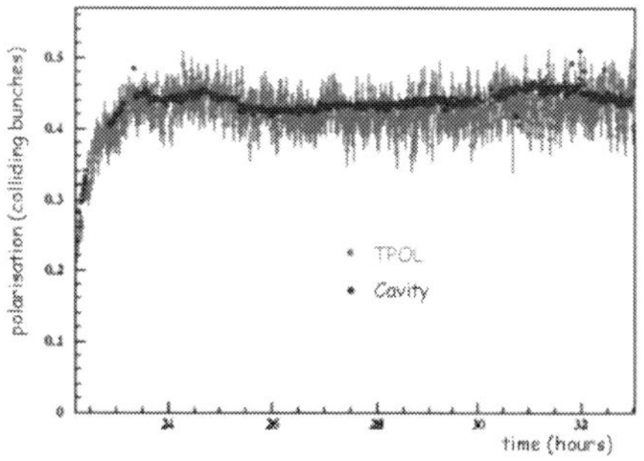

Figure 5. Polarisation measurement from the LPOL cavity (blue points) and the TPOL (red points) for the colliding bunches as a function of time for about ten hours.

Detailed systematic studies have been performed to determine the total error on P_z. Most of them can be estimated from the data themselves by using the full LPOL cavity data set and including also dedicated data taking periods with non-standard setups [7]. An overall relative systematic error of 0.9% is quoted on P_z, this number including all the systematics except the knowledge of S_3 which cannot be determined from the data themselves.

3. Control of S_3

The uncertainty on the S_3 determination comes mainly from two sources: (1) the measurement of S_3 itself performed with an optical system usually located close to, but outside, the electron beam pipe and, (2) the transport of the measured S_3 value through optical elements and vacuum window up to the electron-laser IP. The method to measure S_3 and the transport of the measured value are descibed in this section.

A schematic overview of the Fabry-Perot cavity optical setup is presented in figure 6. The laser beam passes first through a Glan-Thomson prism in order to provide a purely linearly polarised state and then through a quater wave plate (labeled $\mathrm{QWP_{ent}}$) to provide an elliptical polarised state. The beam then passes through the entrance optics which is composed of a glass plate (used to pick up a fraction of the beam for the locking procedure exploiting the "Pound-Drever" technique [13]) and two lenses (used to match the laser beam to the cavity fundamental mode). The beam is then precisely aligned with four mirrors (of which two are motorised) before entering the two meter long Fabry-Perot cavity.

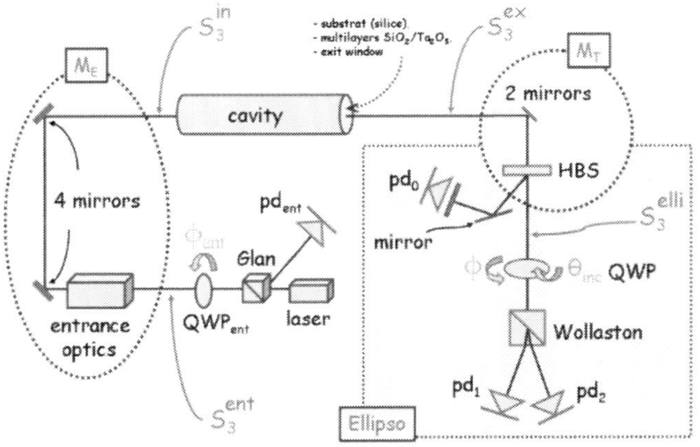

Figure 6. A schematic view of the Fabry-Perot cavity optical system. The location of four determination points of S_3 is indicated by S_3^{ent}, S_3^{in}, S_3^{ex} and S_3^{elli}. M_E and M_T are two transfer matrices discussed in the following.

At the exit of the cavity, the beam is guided with two mirrors to enter an ellipsometer (labeled "Ellipso") which is the key component in the determination of S_3. The principle of an ellipsometer is to send a light beam, of any unknown polarisation S_3^{elli}, through a quater wave plate (QWP). By rotating the plate, the polarisation state of the light is modified and the state at the exit of the plate depends on the state at the entrance. A polariser (Wollaston prism) placed behind the plate spatially separates the beam into two orthogonal linearly polarised states. The analysis of the intensities of these two beams in the photo-detectors $\mathrm{pd_1}$ and $\mathrm{pd_2}$, for various azimuthal angles ϕ of the QWP, allows the deduction of S_3^{elli}. Before entering the ellipsometer QWP, the beam passes first through a holographic beam sampler (HBS) in order to extract a small fraction of the entrance power (about 1% of the incident beam) measured in the photo-detector $\mathrm{pd_0}$ and used as a reference intensity to compensate the effects due to possible laser power variations.

3.1. Ellipsometer characterisation

The measurement of S_3 with an uncertainty of a few per mill requires a precise control of all the ellipsometer optical components (HBS, Wollaston, photo-detectors, QWP). The birefringence of the HBS alone has been measured before its installation in the cavity system and the result is compatible with zero [6, 14]. The extinction rate of the Wollaston prism has been measured experimentally to be less than a few 10^{-5}. Careful studies of the detection system has shown that the photo-detector intensities $I_{1,2} = i_{pd_{1,2}}/i_{pd_0}$ are known at the per mill level [6, 14]. Since an optical model is needed to reconstruct S_3 from the photometric measurements performed after the polariser, the model accuracy has to be controlled below the per mill level. The QWP is a crucial component of the ellipsometer. It is usually anti-reflection coated with double layers and thus taken as a simple delay plate in basic optical models [15]. However, the reflectance of such coated plates is typically of the order of 0.5%, thus limiting the model accuracy to the same level. In order to decrease the model uncertainty, we followed the work of [16] by choosing an uncoated quartz QWP of high optical quality (delay and thickness tolerances of 1/300 and a few micrometers respectively). In doing so we have to account for multiple reflections inside the anisotropic uniaxial QWP, to model the plate defects and the experimental misalignments, and to perform a fine calibration of the ellipsometer and in particular of the plate thickness (a error of one micrometer on the plate thickness leads to a systematic error of 0.5% on S_3 [6, 14]). The ellipsometer will be first used as a calibration system to characterise precisely some of its optical components. All the details of the model describing the ellipsometer can be found in [6, 14] where the modeling and the calculations relative to the Wollaston cube, the QWP and the optical misalignments are explicitly given.

Two independent ellipsometer calibrations have been performed: one in the clean optical laboratory where the room temperature is regulated at 25° and the other in the HERA tunnel where the room temperature is regulated at 35°. For each calibration, ellipsometer measurements are recorded at various incident angles θ_{inc} between the laser beam and the QWP normal direction, various azimuthal angles ϕ of the QWP, and various initial polarisation state S_3^{elli} (see figure 6). A comparison between the experimental and calculated photo-detectors intensities is performed using a χ^2 minimisation fit. The unknown parameters of this fit are the thickness of the QWP, the quartz birefringence, the light polarisation S_3^{elli}, and all the possible misalignment parameters of the system. The excellent agreement between experimental intensities and theoretical ones based on the fit is illustrated by a typical example in figure 7, where the quantities $I_{1,2}$ are presented for a sample as a function of the azimuthal angle ϕ of the ellipsometer QWP.

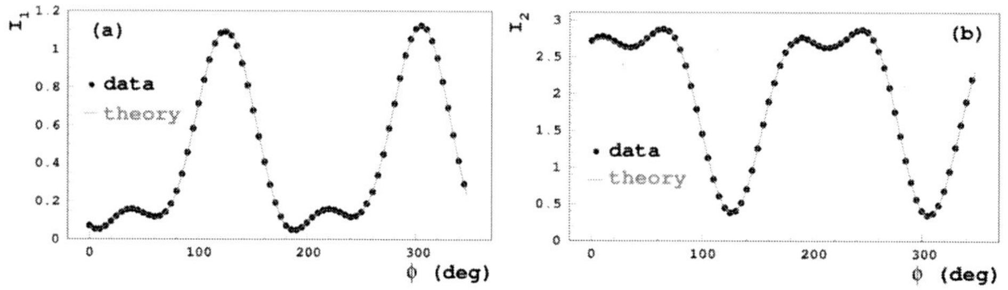

Figure 7. Experimental intensities (black bullets, for clarity only a subsample is shown) I_1 (a) and I_2 (b) compared with theoretical ones (curves) derived from the fit, as a function of the azimuthal angle ϕ of the ellipsometer QWP (for a angle θ_{inc} and a polarisation S_3^{elli} fixed).

All ellipsometer parameters determined from the two independant calibrations are found to be realistic and well defined within a precision of 0.3% [6, 14]. Among them, one interesting quantity is the quartz birefringence value which can be compared with textbook values previously determined at a temperature of 18° and 22° in [17, 18]. Based on the relation of optical index variation with temperature [19], quartz birefringence values of [17, 18] are scaled to $T = 25°$ and at $T = 35°$ and are shown in figure 8 together with the two birefringence values obtained from our fits. Our results agree at better than one per mill with the ones quoted in the references. The determination of the quartz birefringence by the ellipsometer does not aim for the same level of precision as was obtained in [17, 18] but provides a good test of the validity of the model.

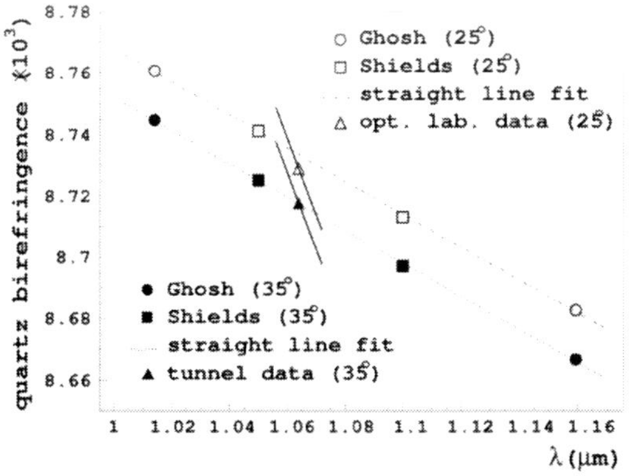

Figure 8. Quartz birefringence values scaled to 25° and 35° from textbook values [17, 18], as a function of the wavelength. Dashed lines are straight line fits to four scaled textbook points. Birefringence values determined from the fits to our two independant data samples are indicated by an open and a solid triangle, respectively. The error bars are inclined for clarity.

3.2. Regular measurements of S_3 and systematics studies

During the data taking period of the LPOL cavity, values of S_3^{elli} are regularly determined. For this, the azimuthal angle of the entrance plate $\mathrm{QWP_{ent}}$ is such that the light is close to a fully right or left circularly polarised state $S_3 = \pm 1$ [7]. Each value of S_3^{elli} is extracted from a data sample recorded with the ellipsometer at various azymuthal angle of the ellipsometer QWP. To extract S_3^{elli}, a fit using the model previously described is performed where the only free parameter is S_3^{elli}, all other parameters remaining fixed to values previously determined by the characterisation of the ellipsometer in the HERA tunnel.

These values S_3^{elli} and their uncertainties of 0.3% quoted in section 3.1 are measurements of the degree of circular polarisation at the entrance of the ellipsometer. What we are interested in is, however, the S_3 value at the electron-laser IP, *i.e.* at the center of the Fabry-Perot cavity. A priori, these two values are the same, but because of the presence of optical components between the two (the exit mirror and the exit window of the Fabry-Perot cavity and the optical system M_T, see figure 6), a small difference could be induced because of a parasitic birefringence. The transport of S_3 have therefore to be studied.

The birefringence of the substrate, the coating and the mounting system of the exit mirror and the exit window has been estimated or measured [12, 6]: the bias induced on S_3 from the center to the exit of the cavity is at the utmost of the order of 3×10^{-5}. The determine the parasitic ellipticity associated to the 45° dielectric mirrors used to guide the light into the ellipsometer, the transfer matrix M_T of the system has been determined and used to correct S_3^{elli} [6, 14]. The correction is calculable for each value of S_3^{elli} and therefore does not enter as a systematic error but is explicitly determined to correct each S_3^{elli}. The corrections are all below five per mill.

Although the previous studies have provided values of S_3 at the electron-laser IP with an uncertainty around three per mill, we have also characterised the entrance optical elements by a matrix M_E and determined the values of S_3 at the four different locations S_3^{ent}, S_3^{in}, S_3^{ex} and S_3^{elli}

indicated in Fig. 6 in order to check their coherence and make sure that no additional unknown large effect could induce a bias on S_3 at the center of the cavity. This study has shown that S_3 is controlled along the whole optical path at the five per mill level [6, 14].

4. Conclusion

A new Fabry-Perot cavity polarimeter has been constructed and successfully operated at HERA. The measurement of the electron beam polarisation is reported for the first time in the few photon mode leading to a statistical precision of 2% per bunch per minute. Detailed systematic studies have been performed resulting in a total relative systematic uncertainty of about 1%, which is a factor of $2-3$ smaller than the precision quoted currently by the other polarimeters at HERA. To reach such a small systematic uncertainty, on one hand we have used the possibility to describe the few photon energy spectra from first principles by convoluting the signal and background QED processes with the detector effects, and on the other hand we have determined the degree of circular polarisation S_3 of the laser at the electron-laser IP with an uncertainy of 0.3%. The level of accuracy presented here has, to our knowledge, never been reached in the environment of a particle collider and provides a good prospect for applications in a future linear collider [20, 21, 22].

References

[1] Beckmann M *et al* 2002 *Nucl. Instr. Meth.* A **479** 334
[2] Barber D P *et al* 1993 *Nucl. Instr. Meth.* A **329** 79
[3] Falletto N *et al* 2001 *Nucl. Instr. Meth.* A **459** 412
[4] Baylac M *et al* 2002 *Phys. Lett.* B **539** 8
[5] Escoffier S *et al* 2001 *Nucl. Instr. Meth.* A **551** 563
[6] Jacquet M 2009 *Habilitation thesis* LAL 09-210
[7] Baudrand S *et al* 2010 *J. Instr.* JINST5 P06005
[8] Piotrzkowski K 1995 *Z. Phys.* C **67** 577
[9] Tsai Y 1974 *Rev. Mod. Phys.* **46** 815
[10] Lomperski M 1993 *Internal Report* DESY 93-045
[11] Lorenzon W *et al* 1995 *Internal Report* HERMES 95-023
[12] Zomer F 2003 *Habilitation thesis* LAL 03-12
[13] Black E D 2001 *Am. J. Phys.* **69** 79
[14] Brisson V *et al* 2010 *J. Instr* JINST5 P06006
[15] Clark Jones R 1941 *J. Opt. Soc. Am.* **31** 488
[16] Poirson J, Lanternier T, Cotteverte J C, Le Floch A and Bretenaker F 1995 *App. Opt.* **34** 6806
[17] Ghosh G 1999 *Opt. Com.* **163** 95
[18] Shields J H and Ellis J W 1956 *J. Opt. Soc. Am.* **46** 263
[19] Toyoda T and Yabe M 1983 *J. Phys.* D: *App. Phys.* **16** 97
[20] Long K and Schüler K P 2002 *Nucl. Instr. Meth.* A **494** 75
[21] Schüler K P 2001 *Proc. of the Linear Collider Workshop 2000 (Batavia)* (Batavia 2000, Physics and experiments with future linear e^+e^- colliders) p 744
[22] Moortgat-Pick G *et al* 2008 *Phys. Rep.* **460** 131

Workshop on Sources of Polarized Leptons and High Brightness Electron Beams (PESP2010) IOP Publishing
Journal of Physics: Conference Series **298** (2011) 012024 doi:10.1088/1742-6596/298/1/012024

Vector polarimetry at MaMi - Measurements of tensor correlation coefficients in e$^-$ bremsstrahlungs processes

F. Nillius and K. Aulenbacher

Institut für Kernphysik Johannes Gutenberg-Universität Mainz

E-mail: `nillius@kph.uni-mainz.de`

Abstract. Electron/photon tensor-correlation coefficients may allow to design a polarimeter that can measure all components of beam polarization simultaneously—a so-called *vector polarimeter*. Besides its purpose as a beam diagnostic device this vector polarimeter would also allow to test theoretical predictions for the electron-photon polarization correlations at energies between 1 MeV and 3.5 MeV. As a first step we have set up a measurement of the helicity transfer to the photon as a function of energy which is based on the Compton absorption method.

1. Introduction

Since the beam polarization is a vector, it is interesting to measure all its polarization components simultaneously[1]. An already existing polarimeter at MAMI [2] is based on elastic electron Mott scattering at an electron energy of 3.5 MeV. It is only able to measure the component of beam polarization which is perpendicular to its scattering plane, defined by the direction of the incident electron beam and the scattered electron. It is possible to measure the second transverse component by installing another Mott detector with a detection plane that is rotated by 90 degree. However, this approach requires considerable effort due to space restrictions under the extreme backward angles required. The longitudinal component is not accessible at all for the Mott polarimeter, since the scattering asymmetry is zero, due to parity conservation.

An alternative approach is to analyze the polarization of bremsstrahlung photons. The outgoing photon then defines the scattering plane together with the incoming electron beam. The correlation between the polarization components of incoming electron and the outgoing photon is given by a tensor C_{ij} introduced by [3] with $i, j = 0, 1, 2, 3$. The first index stands for the electron components

0 = intensity,
1 = transverse polarization in scattering plane,
2 = transverse polarization perpendicular to scattering plane and
3 = longitudinal polarization.

[1] It is of course not possible to measure more than one component of the spin of a single electron. Here, we analyze an electron beam by detecting scattering events from different electrons, which allows to define the polarization vector as an expectation value over the ensemble, see [1] page 12 for details.

The second index for the photon components is chosen as the Stokes parameters:

0 = intensity,
1 = linear diagonal polarization (45°/135°),
2 = circular polarization and
3 = linear polarization in horizontal or vertical direction.

Two out of the eight non-zero components of C_{ij}, namely C_{20} and C_{32}, are of particular interest for our purposes. After a brief discussion of C_{20} we will focus on C_{32} throughout this paper. The coefficient C_{20} describes a correlation of photon intensity with transverse electron polarization. Measuring the intensity for opposite transverse electron polarisations leads to an intensity asymmetry. This observable is similar to the Mott asymmetry and therefore allows to measure two transverse polarization components if two detection systems are operated with scattering planes perpendicular to each other. Calculations by Tseng [4] show that the absolute value of C_{20} is comparable with the Mott analyzing power for MeV energies.

It remains to be demonstrated which principle (Mott or C_{20} gamma asymmetry) is more suitable, but it is evident that the detection of the transverse components is possible. The remaining longitudinal component of electron polarization can be measured by utilizing the coefficient C_{32} which stands for the generation of circularly polarized gamma radiation from the longitudinal electron polarization component. The transfer of helicity from longitudinally polarized electrons to circular polarized photons was calculated by Olsen and Maximon for ultra-relativistic electrons [5]. The helicity transfer according to this approximation is shown as a function of photon energy in figure 1. It is well known that the circular polarization of γ rays can be measured by spin dependent absorption in magnetized iron [6]. In a second paper within these proceedings [7] it is demonstrated that this absorption technique is especially effective in an energy range between 2.5 MeV and 10 MeV. Even if the intensity of the *integral* of the photon flux is measured, asymmetries of the order of 1% can be observed. Together with the very high gamma flux in forward direction this creates ideal conditions for fast and precise integral (i.e. not energy resolved) measurements.

The correlation coefficients are depending on beam and photon energy, the scattering angle and on the nuclear charge. In the intermediate (i.e. not 'ultra'relativistic) energy range, more refined theoretical calculations [4] show significant deviations from the values predicted by Olsen. To our knowledge, almost no measurements of such correlation coefficients have been done in the multi MeV range. The considerations for a vector polarimeter based on detection of bremsstrahlung therefore also lead to the possibility to test these theoretical predictions. We present here an experiment exploring the possibilities for a measurement of C_{32} in the forward direction at 3.5 MeV. The gamma radiation was produced on a lead target (Z=82).

1.1. Calculation of the transmission asymmetry

For longitudinal polarized electrons, the photon flux has no asymmetry in intensity. The bremsstrahlung spectrum $I_0(E_\gamma)$ can be calculated by the Bethe-Heitler formula [8]. This spectrum has an energy dependent degree of circular polarization $\xi(E_\gamma)$ which is proportional to the electron polarization. The circular polarization approaches the electron polarization at the high energy edge of the spectrum (figure 1). Before being detected, the photons have to pass an iron absorber. The main effect of the absorber is to scatter the forward angle emitted photons out of the acceptance of the detector, from the point of the detector this represents an 'absorption'.

The observed attenuation is then $\propto \exp(-(\sigma_0 \pm h\sigma_p)N_A\rho \cdot d/A)$, with ρ and d density and thickness of the absorber and N_A, A Avogadros number and atomic mass of the absorber. For the different helicity states of the photons we have $h = \pm 1$. The cross section σ_0 is the

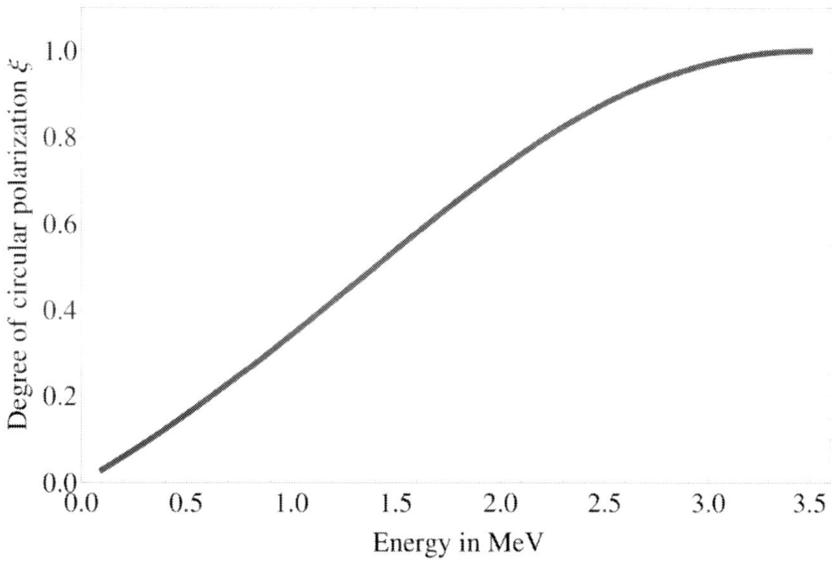

Figure 1. Degree of circular polarization ξ/P_e as a function of photon energy. Electron beam kinetic energy 3.5 MeV.

unpolarized (helicity averaged) cross section - usually dominated by the Compton cross section in our energy range, though photo absorption and pair production also contribute. The cross section σ_p is polarization dependent, which is taken into account by multiplying it with h, the only contributing process at relevant level is Compton scattering[2]. In order to obtain σ_0 the elementary (gamma/e) unpolarized Compton cross section has to be multiplied by the number of electrons per atom (Z), whereas the elementary polarized cross section must be multiplied by the number of *polarized* electrons per atom (typically ≈ 2 in magnetized iron) to obtain σ_p. When switching the longitudinal polarization of the electron beam, the sign of h is flipped. Since the asymmetry is defined as $A = \frac{I(+)-I(-)}{I(+)+I(-)}$ all helicity independent terms cancel and the asymmetry for completely polarized photons is

$$A_{full}(E_\gamma) = \frac{e^{-\sigma_p N_A \rho d/A} - e^{+\sigma_p N_A \rho d/A}}{e^{-\sigma_p N_A \rho d/A} + e^{+\sigma_p N_A \rho d/A}} = -\tanh(\sigma_p N_A \rho d/A). \tag{1}$$

A partially circular polarized photon beam can be described as incoherent superposition of two totally polarized substates of different intensity $I_{(left-circ)} = \frac{1}{2}(1 - P_{circ})$ and $I_{(right-circ)} = \frac{1}{2}(1 + P_{circ})$. The two subspecies will have asymmetries $\pm A_{full}$ which are added to $A = A_{full}(I_{(right-circ)} - I_{(left-circ)}) = P_{circ}A_{full}$. The circular polarization in turn is given by $P_{circ} = P_e C_{32}$ (from the same argument) and therefore the asymmetry should be

$$A\left(E_\gamma\right) = -P_e C_{32}\left(E_\gamma\right) \tanh\left(\sigma_p\left(E_\gamma\right) N_A \rho d/A\right) \approx -P_e C_{32}\left(E_\gamma\right) \sigma_p\left(E_\gamma\right) N_A \rho d/A. \tag{2}$$

The approximation holds if $E_\gamma \sigma_p(E_\gamma) N_A \rho d/A \ll 1$. In order to obtain a prediction for $A\left(E_\gamma\right)$ the theoretical values for $C_{32}(E_\gamma)$ from [5] and the polarization dependent cross section from [6] where used. Besides this, the spin density of the electrons had to be plugged in, for which we have used 8%. The resulting prediction is shown by the blue curve in figure 3.

[2] A more quantitative discussion of σ_0, σ_p can be found in [7].

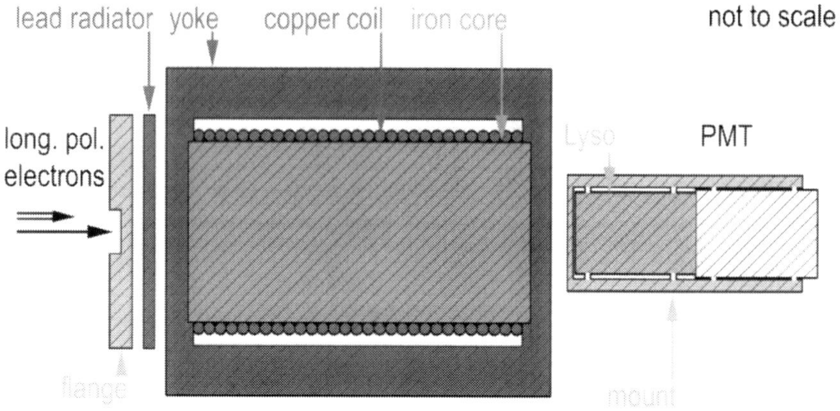

Figure 2. Polarized electrons are transmitted through the end flange and produce radiation on a lead foil. After passing an absorber the remaining photons are detected.

1.2. Principle of the measurement

The experiment took place at the 3.5 MeV beamline of MAMI which is presented in more detail in [7]. The photon absorption experiment was installed in the beam dump of the Mott polarimeter, located about 2 m downstream from it. During the measurements the Mott targets where removed and the beam entered the device shown in figure 2.

The electron beam can leave the vacuum of the accelerator through a thin aluminum window and hits a lead foil of 4 mm thickness. There, the electron beam produces circular polarized photons in bremsstrahlung-processes. Passing a magnetized cylinder the number of photons detected behind this absorber depends on the degree of circular polarization of the photon beam and the degree of magnetization. It is advantageous to measure in a regime where the magnetization of the absorber is saturated. The asymmetry also depends upon the number density of polarized electrons inside the material, so it has to be a material with a high magnetic saturation field.

1.3. Hardware design

- *magnetic absorber:* The magnetic absorber is made of VACOFLUX50 which is a Cobalt-Iron alloy that has a maximum saturation polarization at 2.35 T. The core is 60 mm in diameter and 200 mm long. A copper coil with 500 turns was winded on the surface to excite the core.
- *photon detector:* The photon spectrum is detected with a fast scintillation crystal. LYSO[3] has the advantages of a high density, a fast decay time and good light output and energy resolution.
- *read out electronics:* For read out, the electronic lab of our institute has developed a fast ADC and histogram unit. The pulse identification works with a CFD. The ADC has 12 Bit resolution and data transfer is realized via VME bus.

Taking these components together it is possible to test the hardware setup with a ^{60}Co source. The detector contains the radioactive isotope ^{176}Lu, a naturally occurring β emitter. ^{176}Lu decays via β decays to ^{176}Hf almost exclusively to the 597 keV excited state. This state decays with a three γ ray cascade of 307 keV, 202 keV and 88 keV. This natural spectrum can be used for energy calibration.

[3] Cerium-doped Lutetium Yttrium Orthosilicate

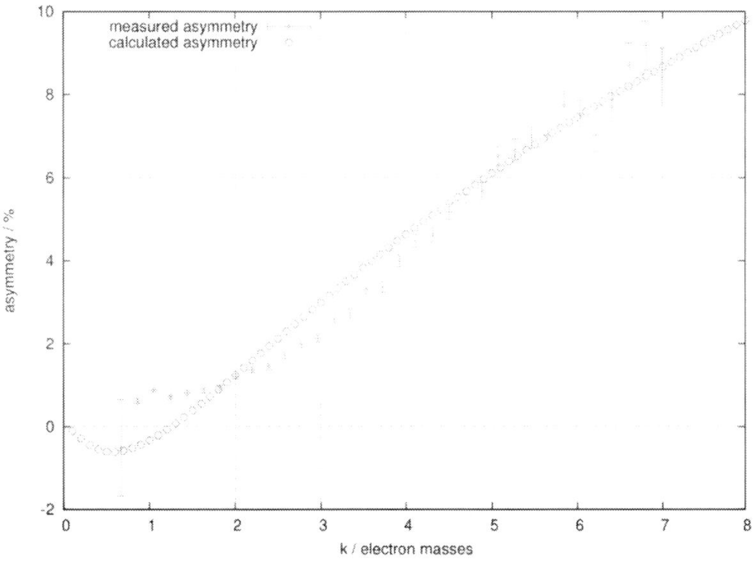

Figure 3. Calculated (blue circles) and measured (red crosses) asymmetry in percent versus photon energy in electron masses. The error bars are statistical errors only.

2. Results

Taking two spectra for opposite beam polarizations, it is possible to determine the asymmetry as a function of photon energy as shown in figure 3. For photon energies higher than four electron masses theoretical prediction and measurements are in agreement. For lower energies considerable discrepancies occur, note especially the predicted sign change of asymmetry at 1.2 electron-masses, which is not observed. We suspect that at low energies a contamination from high energy quanta occurs: Due to the small scintillator size many of these quanta are not fully absorbed in the scintillator and are therefore registered as a low energy signal. The high energy quanta carry a large positive asymmetry and will therefore shift the signal upwards. Such an effect cannot explain the discrepancy with opposite sign in the region between 2 and 4 electron masses, but many explanations, including using a more realistic calculation of C_{32} are possible and have to be considered.

For this reason we cannot yet extract $C_{32}(E_\gamma)$ from the data. A careful consideration of signal diluting backgrounds is necessary, in order to define the systematic errors. The relative size of these errors will increase with decreasing photon energy. Other systematic error sources are for instance the electron beam polarization ($\approx \Delta P_e/P = 2\%$) and uncertainties caused by the absorber magnetization, especially in the end caps of the core, where the magnetization is not complete. Furthermore, a deconvolution of the spectra due to the finite energy resolution has to be performed.

3. Summary & Outlook

We have measured the photon transmission asymmetry $A(E_\gamma)$ which is induced by $C_{32}(E_\gamma)$ for photon energies up to 3.5 MeV by developing a detector system based on LYSO scintillation crystals and fast read out hardware. This detection system will also be used in future measurements of C_{20} to get access to the transverse components of the polarization vector. This will enable the system to work as a vector polarimeter.

Acknowledgments
This work was supported by the Deutsche Forschungsgemeinschaft through SFB 443.

References
[1] Kessler J 1985 *Polarized electrons* Springer series on atoms + plasmas (Springer-Verlag) ISBN 9783540157366 URL http://books.google.de/books?id=pMOy9AMjzd8C
[2] Aulenbacher K and Tioukine V 2009 *AIP Conference Proceedings* **1149** 1155–1159 URL http://link.aip.org/link/?APC/1149/1155/1
[3] Tseng H K and Pratt R H 1973 *Phys. Rev. A* **7** 1502–1515
[4] Tseng H 1999 *Chinese Journal of Physics* **37**
[5] Olsen H and Maximon L C 1959 *Phys. Rev.* **114** 887–904
[6] Schopper H 1958 *Nuclear Instruments* **3** 158 – 176
[7] Barday R 2010 Contribution to this conference
[8] Bethe H and Heitler W 1934 *Proceedings of the Royal Society of London. Series A* **146** 83–112 (*Preprint* http://rspa.royalsocietypublishing.org/content/146/856/83.full.pdf+html) URL http://rspa.royalsocietypublishing.org/content/146/856/83.short

AUTHOR INDEX

AUTHOR INDEX